Colours of the stars

Reflection nebula in Orion.

Colours of the stars

---

by
David Malin
Anglo-Australian Observatory
and
Paul Murdin
Royal Greenwich Observatory

Cambridge University Press
Cambridge London New York New Rochelle
Melbourne Sydney

Published by the Press Syndicate of the University of Cambridge
The Pitt Building, Trumpington Street, Cambridge CB2 1RP
32 East 57th Street, New York, NY 10022, USA
296 Beaconsfield Parade, Middle Park, Melbourne 3206, Australia

First published 1984

Printed in Great Britain

Library of Congress catalogue card number: 83-20928

British Library Cataloguing in Publication Data

Malin, David
Colours of the stars.
1. Stars–Colour
I. Title II. Murdin, Paul
523.8'2 QB816

ISBN 0 521 25714 X

SE

Contents

Preface

As we write this, in the heat of an Australian summer, it is exactly 100 years since Ainslee Common, in the chill of an English winter, obtained the first astronomical pictures which showed objects which had not been seen before. His achievement marked the true beginning of astronomical photography. Nowadays with new emulsions and new techniques, photography, in conjunction with electronic detectors, reveals ever more detailed pictures of ever fainter, more complex and more distant stars, nebulae and galaxies. For their understanding of the menagerie of quasars, black holes, white dwarfs and exploding galaxies, astronomers rely on novel instruments and their jargon buzzes with names and acronyms like 'CCD', 'TAURUS', '25 cm camera of the RGO Spectrograph, with IPCS and 1200 line grating'. But first they reach for a picture to show them where to point their telescope and how to plan their observations. In fact pictures often spark the questions and doubts which yield advances in knowledge. Is this fuzzy spot at the centre of an X-ray error box the X-ray galaxy? Is this quasar at the end of this galaxy's spiral arm connected with it? If anyone doubts the usefulness of photography in astronomy, let him participate in infrared astronomy nowadays. As David Allen has remarked to us, with no imaging device available, pictures in the infrared are assembled laboriously point by point and are crude and unsatisfying. Without imaging techniques how long would it have been before optical astronomers recognised similarities and differences between spiral and elliptical galaxies?

With the advent of colour photography the astronomical image takes on a new depth, and more interrelations between parts of the picture become apparent. This is because the colour in a photograph indicates the underlying spectrum, so painstakingly measured by spectrograph and detector. Colour and photography bear directly on the meaning of an astronomical image. Even the pale colours visible to the unaided eye make astronomical sense.

In this book we show how colour is related to the objects in astronomy, how the colours of stars, nebulae and galaxies can be seen and photographed, and how colour and photography enhance our astronomical understanding.

Astrophotographers have constantly returned to one astronomical nebula for their experiments, repeatedly photographing the first object successfully recorded by Common in 1883. The Orion Nebula figures throughout this book. Each of the photographs tells us something different about the Orion Nebula. The eleven are listed overleaf.

PHOTOGRAPHS IN THIS BOOK OF THE ORION NEBULA

Page	Fig.	Date	Telescope	Process
26	—	1984	Hasselblad	Colour film
29	26	1883	Common's reflector	Early dry gelatin plate
47	38	1974	Crossley reflector	Dye-transfer process
53	42	1979	AAT	Unsharp Mask
62	—	1977	INT	3-Colour addition
108	84	1979	AAT	Unsharp Mask
109	85	1979	AAT	3-Colour addition
110	86	1982	UKST	3-Colour addition
111	87	1979,80	AAT	Special contact copying
112	88*a*	1977	UKST	Unsharp masking
112	88*b*	1979	UKST	Amplification
114	90	1979	Aero Ektar lens	Amplification
182	—	1984	IRAS	IR detectors
Cover	—	1983	AAT	3-Colour addition

Acknowledgements

We want to acknowledge the contribution to this book made by the staffs of the Anglo-Australian Observatory and the UK Schmidt Telescope Unit of the Royal Observatory, Edinburgh. Their efforts in successfully operating the Anglo-Australian Telescope and the Schmidt Telescope made possible many of the photographs reproduced in this book.

We also owe thanks to numerous individuals who have helped us:

Dave Hanes, Dave Allen and Lesley Murdin for critically reading parts or all of the drafts;

Phillipa Malin and Beryl Andrews for secretarial help;

Robyn Bland and Janet Dudley, librarians at the AAO and RGO, respectively, for bibliographical and archival searches;

Keith Tritton, who, while Officer-in-Charge of UKSTU, originated its colour photography programme;

Joe Wampler and Don Morton, former and present directors of the AAO, who encouraged Malin in the colour photography project;

the British and Australian astronomers who have given us access to original plates taken for their own purposes, so that we could make colour pictures from them.

Figs 2, 15, 17, 19, 24, 35, 37, 40, 41, 42, 44, 50, 54, 57, 59, 61, 63, 64, 75, 76, 77, 78, 79, 81, 84, 85, 87, 90, 92, 93, 95, 96, 98, 99, 100, 102, 113, 114, 115, 116, 120, 122b, 123, 125, 128, 129, 130, 132, 133, 134, 136, 141, and on p. 106 are copyright © Anglo-Australian Telescope Board 1977, 1978, 1979, 1980, 1981, 1982, 1983.

Figs 28, 36, 43, 45, 46, 47, 48, 56, 80, 83, 86, 88, 89, 91, 92, 101, 112, 117, 118, 119, 121, 122a, 124, 131, 135, 137, 138, 139, 140, and on p. 126 are copyright © Royal Observatory, Edinburgh, 1977, 1979, 1980, 1981, 1982, 1983.

Fig. 3 was created by Bob Fosbury and Ken Hartley.

Fig. 7 is courtesy of Kitt Peak National Observatory.

Fig. 26 was made by D. Calvert from prints from Common's pictures supplied by the Royal Astronomical Society.

Fig. 27 is courtesy of Eastman Kodak Company.

Fig. 31 is courtesy of Ciba-Geigy S.A.

Figs 32, 65 and 126 are copyright © California Institute of Technology, 1958, 1960; for these we are grateful to R.J. Brucato for access to Bill Miller's original photographs.

Figs 33, 71 and 124 are copyright © California Institute of Technology 1955, 1958.

Fig. 37 is courtesy of Gall and Inglis.

Fig. 38 is a Lick Observatory Photograph.

Fig. 49 uses a photo, courtesy of Kitt Peak National Observatory.

Figs 52 and 53 are courtesy of Lund Observatory.

Fig. 62 was taken by Steve Lee.

Fig. 69 was made by Serge Koutchny.

Fig. 70 was taken by Goran Scharmer.

Fig. 105 and the pictures on p. 62 and p. 92 are by David Calvert.

Fig. 108 is from NASA-GSFC-JPL.

Fig. 143 is a NASA photograph.

Chapter One

Colours of the stars

Colour in astronomy

Black and white television is comprehensible and yields the vast majority of the content of a drama or news programme. Colour is therefore not generally necessary to the perception of a scene – indeed those individuals who are colour-blind manage perfectly well except in highly specialised circumstances, such as becoming a pilot or an engine driver; but those who remember changing from a monochrome TV to a colour TV, or who view black and white TV after being used to a colour receiver will realise that colour does contain extra information. Perhaps the information is trivial, such as the colour of the lead actress's gown. Perhaps the information is desirable, such as the colour of the shirts of two opposing football teams. Certainly colour adds an extra dimension to our perception.

Some fields of natural history have come to rely to a degree on colour in order to classify or distinguish – it is at first hard to imagine identifying a flower, bird or animal without colour information, although on second thoughts we would realise that the correct species name is based on more objective criteria such as the number, size and shape of petals, bones or teeth.

In the formal science of astronomy colour plays little part. The colour that can be seen with the eye is subtle; it is weak evidence in the study of the stars, nebulae or planets. The telescopic view of a planet is remarkable for the paleness of the colour. Mars can crudely be said to be red with green markings, but the English vocabulary contains no words which convey the pale pink-orange or faded grey-green which underly this brief shorthand. Jupiter seems coloured with numerous hues of pink, orange, brown, red and even blue, but in tone these colours are those of a thoroughly washed water-colour.

Amateur astronomers, however, enjoy looking for and arguing about the colours of stars. There is a difference in colour between Sirius and Betelgeuse as they shine brightly on a frosty evening; but one amateur sees Betelgeuse as *orange*, another as *deep red*, another, more poetic, as *garnet*. Double stars show interesting colours, the more readily seen because the two stars are close together and because of the contrast of one colour against the other. Colour is thus present in astronomy. Even in the vocabulary of the professional science of astronomy there is a strong underlying hint of colour. Astronomers talk of the *colour* index of stars. The Sun, they say, is a *yellow* dwarf, Betelgeuse a *red* supergiant. Some stars, they claim, have a *colour* excess and galaxies are *red*-shifted. Galaxies are made, astronomers say, of *blue* Population I and *red* Population II stars. These technical terms, which will be explained later in the book, demonstrate the underlying presence and importance of colour in astronomy.

Astronomers, then, measure colour and use these measurements to classify and distinguish, and colour must be present in stars, the galaxies and the nebulae. Colour *is* present. It needs skill and technique to make it apparent, but the Universe *is* a coloured one. It cannot be seen easily by the unaided eye for two reasons, one intrinsic to astronomy and one within the physiology of our eyes, but with photographic techniques we can overcome both limitations. The colours revealed are true to reality, as much as the coloured scenes on a television receiver are true to the reality before the camera. The colours

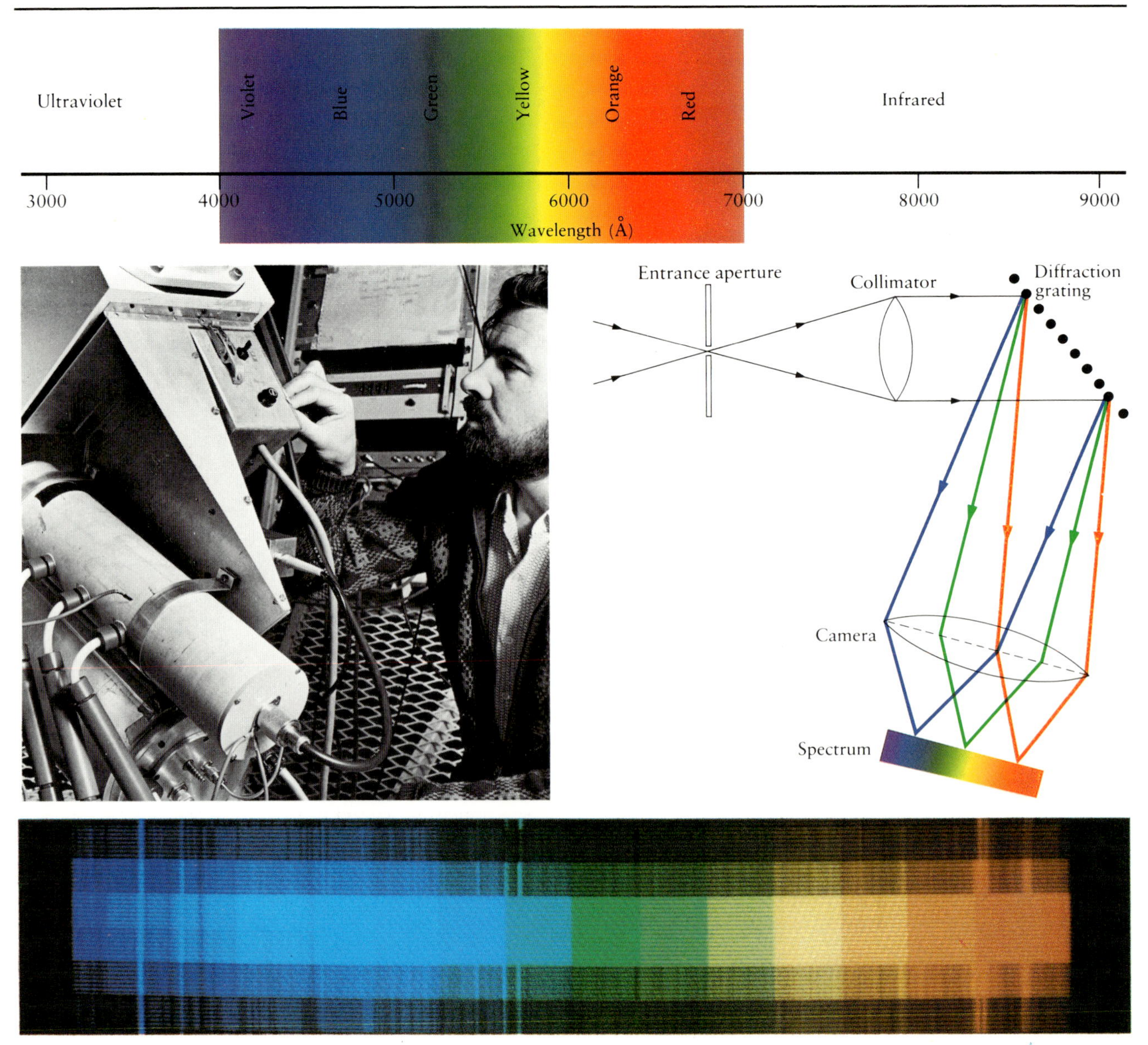

Fig. 1. *(top) Colour and the wavelength of light. The visible band of light runs from about 4000 to about 7000 Å Ultraviolet radiation and infrared radiation lie outside this visible band. Within the visible light band runs the gradation of the colours of the rainbow from violet to red. Although there are actually no distinct boundaries, violet, blue, green, yellow, orange and red occupy roughly equal 500 Å steps from 4000 to 7000 Å.*

Fig. 2. *(centre) Spectrograph and how it works. (left) An engineer adjusts an astronomical spectrograph and its electronic detector as it hangs below a large telescope. (right) Starlight enters a slit-shaped entrance aperture, is rendered parallel by a collimator, and is dispersed by a diffraction grating into a spectrum which is imaged by a camera on to the detector.*

Fig. 3. *(bottom) The spectrum of the nucleus of a galaxy called PKS 2152–380, recorded by the Anglo-Australian Telescope and displayed on a TV colour monitor of the STARLINK astronomical image processing computer system. The continuum spectrum of the nucleus is represented by the band of colour from ultraviolet to red which has been simulated in the colour display by short bands of colour rather than by the continuous gradation which would be perceived in reality. The ultraviolet end of the spectrum shown here as violet would not in fact be visible to the eye. The nucleus is embedded in a spiral galaxy which contains nebulae whose spectra are emission lines, shown crossing the continuum from top to bottom, i.e. across the galactic nucleus. The violet-coded line on the left is the ultraviolet line at 3727 Å due to ionised oxygen. The strongest red line is Hα (6563 Å), with ionised nitrogen and sulphur lines nearby. Just left of centre are three blue-green lines which are Hβ (4861 Å) and the pair of lines at 4959 and 5007 Å from twice-ionised oxygen.*

are not identical to the colours you would see if you were present, but they are a representation of what exists and they reveal interrelations and convey a vividness which would otherwise pass you by.

To explain what is revealed in the colour pictures in this book, we have to lay some ground work on the nature of colour and of the stars, and this is the subject matter of the rest of this chapter. In Chapter 2 we describe the techniques by which the pictures were created. In the body of the book we show the colour of the stars and describe what the pictures reveal.

Colour and the spectrum

Colour represents the balance of light in a spectrum. White is the appearance of light which comes from something which emits electromagnetic radiation uniformly over the whole visible spectrum. When passed through a suitable medium, such as a glass prism or a drop of rain, white light can be spread into its rainbow-like spectrum of graded colours: red, orange, yellow, green, blue and violet. This sequence of so-called spectral colours corresponds to a sequence of wavelengths of light. Visible light is a wave with a separation between crests in the range 0.39 to about 0.70 micrometres (1 micrometre = 1 μm = one thousandth of a millimetre = one millionth of a metre; one thousandth of an inch is 25 μm). Violet light has the shortest wavelength of any visible colour while red light has the longest wavelength. Outside the visible band lie the infrared (longer wavelengths than red) and ultraviolet (shorter wavelengths than violet). There is a correspondence between wavelengths and spectral colours (Fig. 1).

If a band of light lies in a spectrum within a single narrow wavelength range as shown in Fig. 1, the light appears to have the corresponding spectral colour, shaded more or less true according to its exact position within the range. Spectral colours are strong and distinctive. If the spectrum of light spreads over more than one wavelength range, however, then the colour of the light is not that of a pure spectral colour but is mixed and less distinctive. A lamp may for instance encompass the whole of the visible range but emphasise the longer wavelengths – it will appear to give a yellowish shade. On the other hand, if the shorter wavelengths are emphasised the lamp will appear to give a light blue shade. If the middle wavelengths are de-emphasised the colour of the lamp has a washed-out pinkish appearance.

In astronomy, stars, nebulae and galaxies are perceived by the light they emit, and analysing the spectrum of the light is the main way in which astronomers draw conclusions about the nature of celestial objects. Light from the celestial object which is to be studied is caught in a telescope and funnelled into a device (a spectrograph) for dispersing the light into its spectrum (Fig. 2). The spectrum can be recorded electronically and displayed as a plot of light intensity over a spread of wavelengths. The spectrograph is the astronomer's principal aid in unravelling the story of the light from the stars and nebula.

If a spectrum (Fig. 3) consists of a single isolated wavelength (or relatively few isolated wavelengths) it is said to be an *emission-line spectrum*. Emission-line spectra come from excited atoms in gaseous material and we will see (Chapter 5) that clouds of gaseous material with such spectra (*emission-line nebulae*) are boldly coloured.

A few prominent emission lines are listed in Table 1 with their colour. Perhaps the best-known terrestrial emission lines are those in sodium lamps used for street lighting. The wavelengths of emission lines in astronomy are usually expressed in Angstrom units (Å) (10 000 Å = 1 μm) and the deep yellow colour of sodium lamps is a result of the emission lines at 5890 Å. Other emission lines are produced when gases at low pressure are excited by the flow of an electric current. This process is used to create the vivid colours of advertising signs, and the bright colours of Piccadilly Circus, King's Cross and Times Square are characteristic of emission-line spectra.

On the other hand a spectrum may consist of a broad slew of wavelengths and is then called a *continuous spectrum* (Fig. 3). The colour of light from such a spectrum is less distinctive than that of the spectral colours (Fig. 4). It is a generally weaker pastel shade, much more subtle because of the complex response which it evokes in our brains.

The stars have continuous spectra of this sort and this is the reason why they are not very strongly coloured. The effect of adding a wider range of spectral colours into a spectrum is to make a less distinct, more washed-out, whiter colour, unless one or two of the spectral colours are very strong. This is illustrated by the difference in colour between low-pressure and high-pressure sodium lamps (Fig. 5).

TABLE 1. SOME COMMON SPECTRAL LINES

Colour	Wavelength (Å)	Source
Ultraviolet	3727	Ionised oxygen in rarefied astronomical nebulae
Blue	4358	Mercury lamps
Orange	5889	Sodium lamps
Green	5577	Oxygen atoms in the aurora
Green	5650 }	Light-emitting diodes as found in calculator and clock displays
Red	6350 }	
Red	6300	Oxygen atoms in the aurora

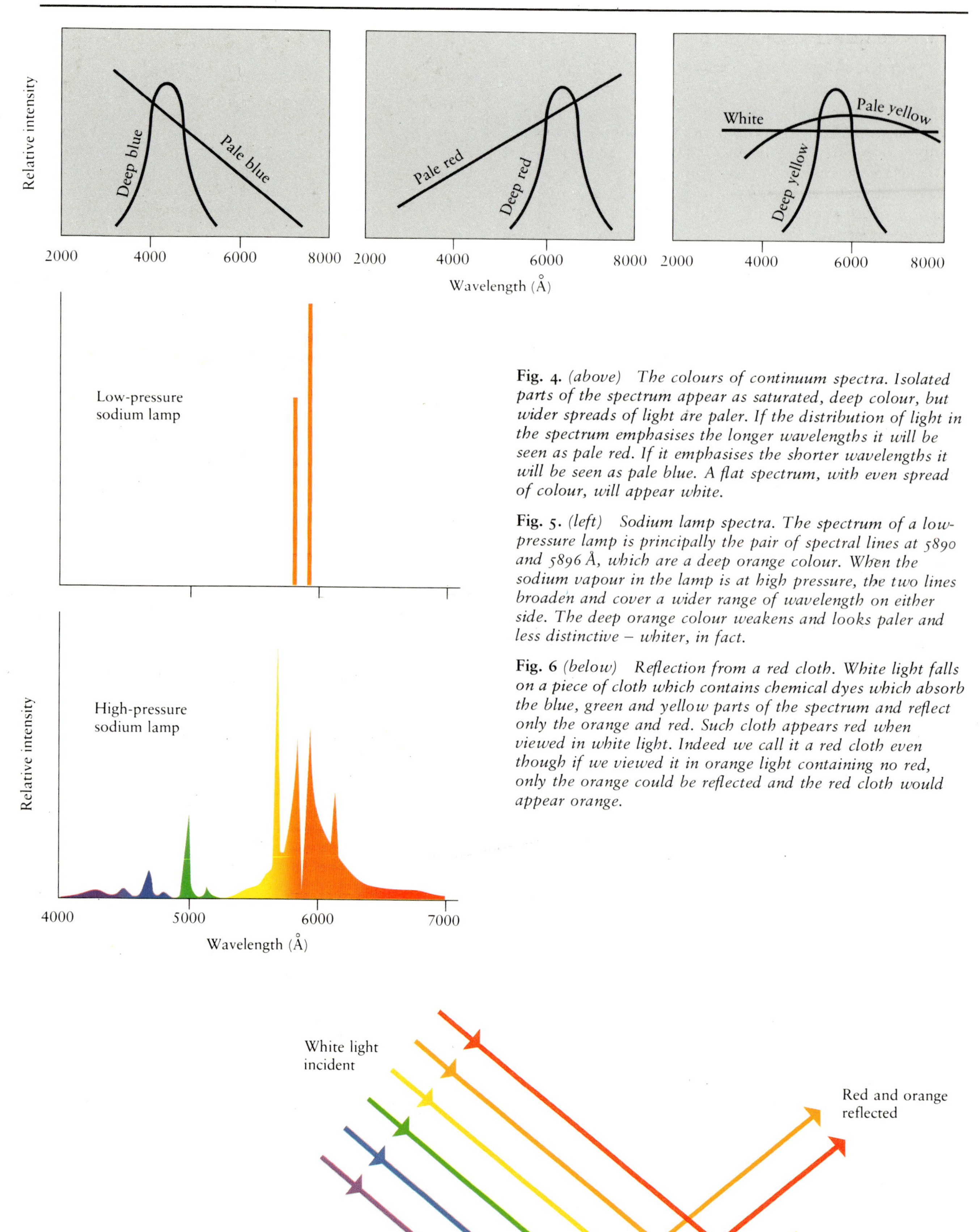

Fig. 4. *(above) The colours of continuum spectra. Isolated parts of the spectrum appear as saturated, deep colour, but wider spreads of light are paler. If the distribution of light in the spectrum emphasises the longer wavelengths it will be seen as pale red. If it emphasises the shorter wavelengths it will be seen as pale blue. A flat spectrum, with even spread of colour, will appear white.*

Fig. 5. *(left) Sodium lamp spectra. The spectrum of a low-pressure lamp is principally the pair of spectral lines at 5890 and 5896 Å, which are a deep orange colour. When the sodium vapour in the lamp is at high pressure, the two lines broaden and cover a wider range of wavelength on either side. The deep orange colour weakens and looks paler and less distinctive – whiter, in fact.*

Fig. 6 *(below) Reflection from a red cloth. White light falls on a piece of cloth which contains chemical dyes which absorb the blue, green and yellow parts of the spectrum and reflect only the orange and red. Such cloth appears red when viewed in white light. Indeed we call it a red cloth even though if we viewed it in orange light containing no red, only the orange could be reflected and the red cloth would appear orange.*

Reflection colours

The balance of light in a spectrum can be altered when light is reflected, and indeed in our daily lives this is how most of the colour of our world is created. Dyes and pigments in the flowers, textiles and surfaces around us selectively absorb parts of the daylight spectrum. What is reflected we perceive as colour. A surface is green, for example, because it absorbs blue and red light; it is yellow because it absorbs only blue. The appearance of the colour of a surface depends on its nature *and* the incident spectrum of light (Fig. 6). This is the reason why art galleries are illuminated best by sky-lights (the natural light which mimics the cold north light of the garret in which paintings are traditionally created) and why people appear peculiar when seen in low-pressure sodium lighting, with sallow faces and purple lips. They look more natural when seen by the light of whiter high-pressure sodium lamps, which is one reason why high-pressure lamps are more acceptable and becoming more popular than the low-pressure variety.

The range of subtle colours seen by reflection is produced by the variety of dyes and pigments in natural things. These so-called chromophores absorb bands of light in the spectrum. The absorption bands may be quite selective, and may either completely absorb a particular wavelength or not. Thus the reflected spectrum from a coloured surface may never be mimicked in the spectrum of stars, where the changes in intensity at different wavelengths are gradual.

Starlight may be reflected from astronomical nebulae. They also produce strong colours, as we shall see. This is not primarily because the nebulae absorb starlight on some coloured surface but because, like coloured surfaces, they are very selective in the way in which they interact with and reflect starlight. For this reason reflection nebulae provide the most diverse colours in the Universe.

Star temperatures

The continuous spectrum of a star comes from its surface – the Sun's surface is represented by the circular outline which can be viewed as the Sun is setting (Fig. 7) – and most stars have a spherical surface similar to the Sun's. The temperature of this surface is what determines the general distribution of light in a star's continuous spectrum. The hotter stars emit more blue light than cooler stars; these latter emit more red light than the hotter stars (Fig. 8). The pure emission spectrum of hot or cool objects is called a black-body spectrum; it has a characteristic shape, cast into mathematical formulae by Max Planck (1858–1947).

The range of temperatures of stars' surfaces is from hotter than 30 000K to cooler than 2500K. (The K signifies that temperatures here are measured in degrees Kelvin, or Celsius degrees measured from absolute zero, which is $-273\,^{\circ}$C. The freezing point of water is $0\,^{\circ}$C which is equivalent to 273K.) The spectra of light from the surfaces of stars thus have the general shapes of those of black bodies of temperatures in that range.

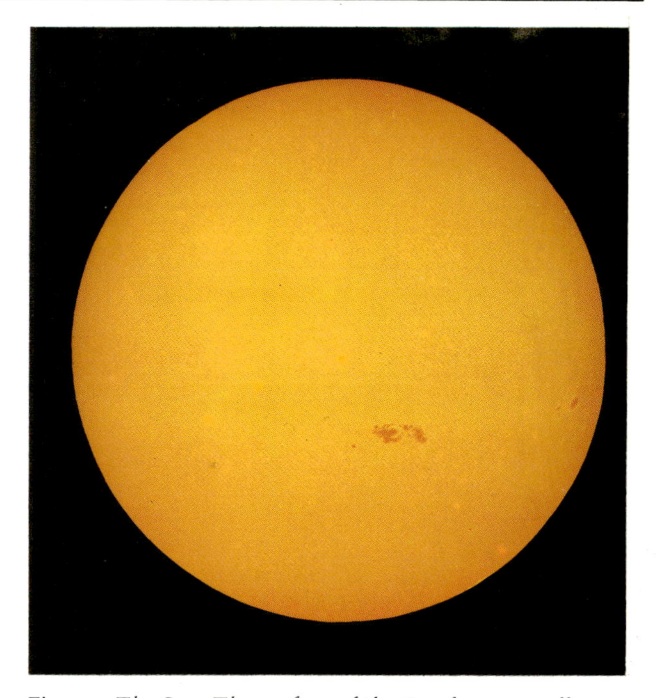

Fig. 7. *The Sun. The surface of the Sun forms a well-defined circular shape, with only slight irregularities at sunspots. The orange of the Sun is represented here with a deeper colour than would be perceived if the Sun was faint enough to be viewed directly.*

Because the progression from cool stars to hot ones is predominantly an orderly sequence of black-body curves, astronomers can determine the temperature of a star by a very simple measure of the balance of light in the spectrum. The ratio of the intensities of light at two arbitrarily chosen wavelengths will give an index of the balance of the light at those wavelengths, its overall colour and hence the star's temperature. To make this judgement astronomers measure the brightness of the star at the two chosen wavelengths.

Star magnitudes

Astronomers, for historical reasons, measure starlight in units called *magnitudes*. Originally the brightest stars which can be seen were called 'stars of the first magnitude' and the dimmest stars which can be seen with the unaided eye are 'sixth magnitude', without there being any understanding of the sizes of the relative brightnesses of such stars. It was empirically determined that the ratio of the brightnesses of stars of first and sixth magnitudes was about 100:1 and so it

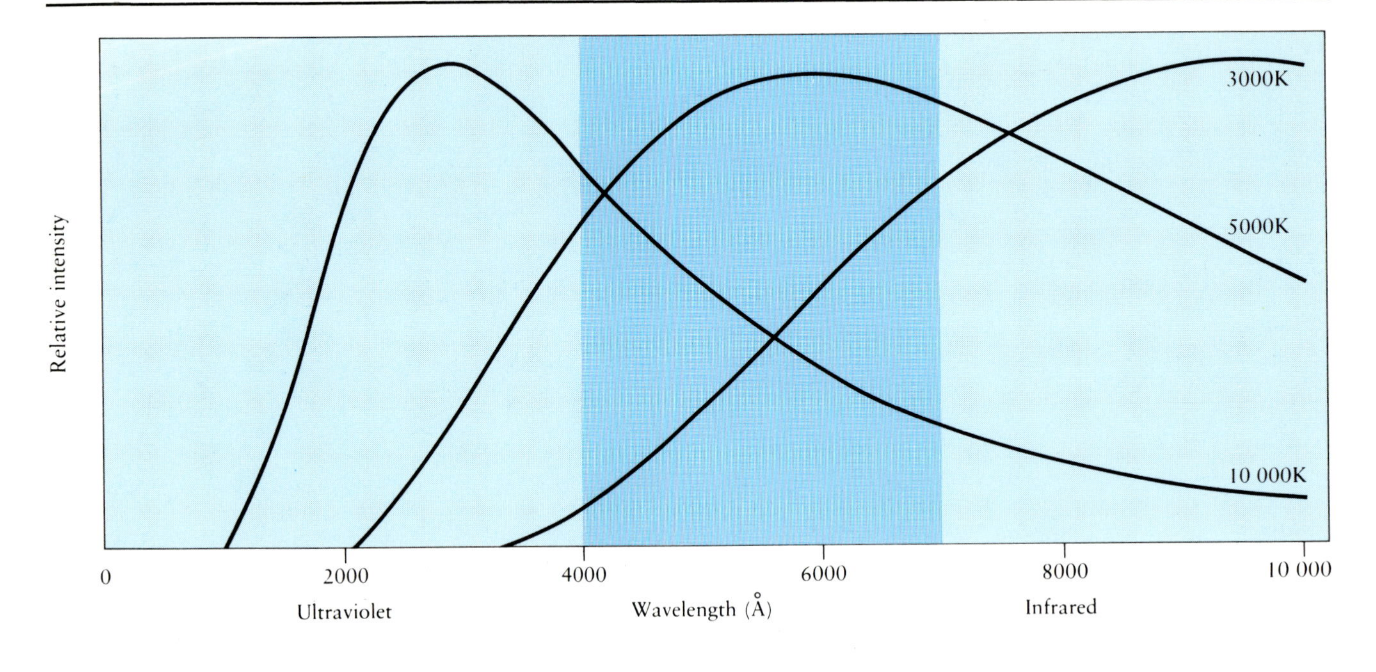

(*a*)

Fig. 8. *Black-body curves.* (a) *(above) The spectra of three stars are shown here, labelled with their temperatures. The shape of each black-body curve consists of a steep rise from the short-wavelength end of the spectrum, a broad region in which emission is strongest, and a weaker fall in intensity to the long wavelengths. The broad maximum in the Planck black-body curve means that most stars (like the one here labelled 5000K) emit a wide range of wavelengths, and are near-white. The cooler stars, like the 3000K black body in which the maximum lies in the infrared, show a strong colour, because the steep fall in output at short wavelengths cuts off the blue and green spectral primary colours and leaves the red, orange and yellow spectral colours to give shades of red and orange to cool stars. In hotter stars, such as the one labelled 10 000K, the maximum of the Planck black-body curve lies in the ultraviolet, but the more gentle fall in intensity from blue to red means that the blue colour of such a star is not pronounced. These curves show only the distribution of light in the spectra of black bodies, not the relative intensities. In fact a hot black body of given surface area radiates at all wavelengths a greater energy than a cool one of the same size.* (b) *(right) The degree to which the black-body curve is a good fit to star spectra can be judged by this comparison of the Sun's spectrum and a black body curve of temperature 5800K. Atoms in the cooler atmosphere of the Sun extract some of the light from the spectrum and the overall continuum spectrum is marked by so-called absorption lines. The effect of the lines on the colour of stars is too small to perceive with the eye but large enough to measure with instruments. The absorption lines lie on the continuum like a blanket, particularly in the blue and ultraviolet end of the spectrum; from this analogy, the effect of the lines on the gross shape of the stars' spectra is called blanketing.*

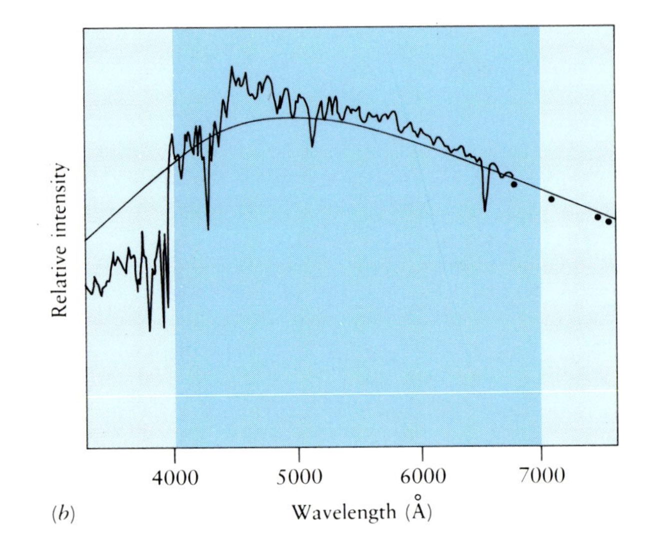

(*b*)

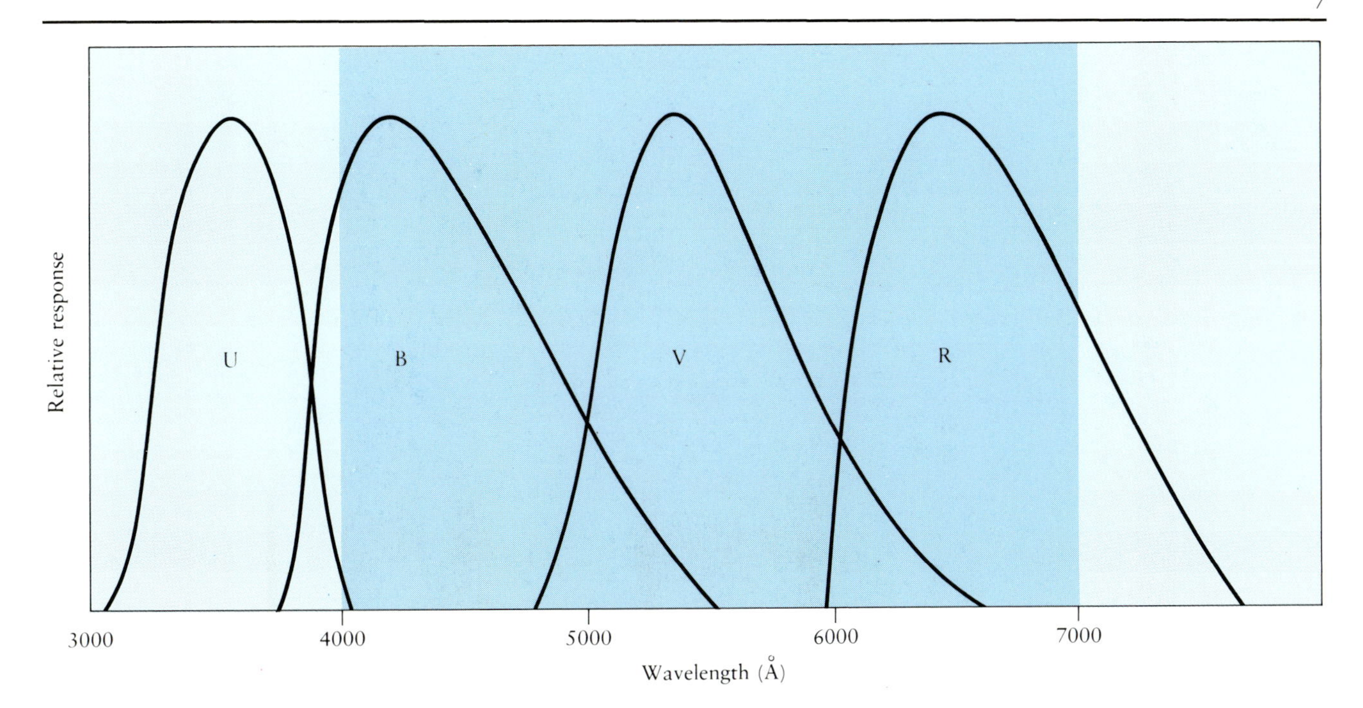

Fig. 9. *Sensitivity of astronomical colour passbands. The UBVR system of photoelectric photometry isolates sections of the spectrum in the ultraviolet, blue, green and red regions, respectively.*

was decided that a five magnitude interval would be defined as exactly this ratio. Thus, a first-magnitude star is 2.512 times as bright as a second-magnitude star which is itself 2.512 times as bright as a third-magnitude star and so on to a sixth-magnitude star, which is in total 100 times fainter than a first-magnitude star ($2.512^5 = 100$). This kind of brightness scale is called a logarithmic scale and the mathematical definition of magnitude, m, is:

$$m = -2.5 \log I + \text{constant}$$

The value of the constant was chosen so that stars which were historically known as first magnitude had a magnitude which was indeed about 1. The intensity, I, in the right-hand side of the equation could be measured in various ways, for example, by eye, or by a photomultiplier, which is an instrument for converting light into electrical impulses which can be counted. Each apparatus has a particular spectral sensitivity, and is thus sensitive to a particular part of the spectrum emitted by the star. Indeed the apparatus can deliberately be made to respond to different parts of the spectrum by covering it with different pieces of coloured glass. A red glass transmits red light, and the intensity of a star measured by a photomultiplier covered with red glass can be converted into the star's 'red magnitude'; the same photomultiplier covered with blue glass yields a 'blue magnitude'. Astronomers have established a standard series of filter-glass/photomultiplier combinations which define so-called magnitude systems. One widely used standard was invented by H.L. Johnson (Fig. 9) and includes pieces of glass which isolate bands of the spectrum in the blue (peak wavelength 0.44μm) and yellow-green (peak wavelength 0.55μm). The latter wavelength corresponds to the peak sensitivity of the eye and this is called the *visual band*. Magnitudes computed on it are symbolised V. The former is the blue band and gives B magnitudes. There are also red magnitudes (R), ultraviolet magnitudes (U), etc.

Colour index

The balance of light in a spectrum (which, as illustrated in Fig. 4, is connected with perceived colour) is measured by the ratio of light intensities in two wavebands, for instance, the blue and visual bands (Fig. 10). Because star magnitudes are logarithmic, the ratio of light intensities transforms to a difference of star magnitudes. The difference, $B-V$, of star magnitudes (or any other two magnitudes in any other pair of wavebands) is called the *colour index* of the star, because it is a numerical measurement of the colour. The larger $B-V$ is, the redder, and cooler the star. Negative $B-V$ indicates that a star tends towards a blue colour. The relation between $B-V$ and temperature for most stars is given in Fig. 11. The reason that this relation exists is that stars' spectra have particular shapes which are very like those of spectra of black bodies.

The concept of using the difference between two colours as an indicator of temperature is quite familiar to photographers. A colour-temperature meter is a device which contains two photocells, one sensitive to blue light, the other to red. The ratio of their outputs (i.e. $B-R$) is indicated on a continuous scale

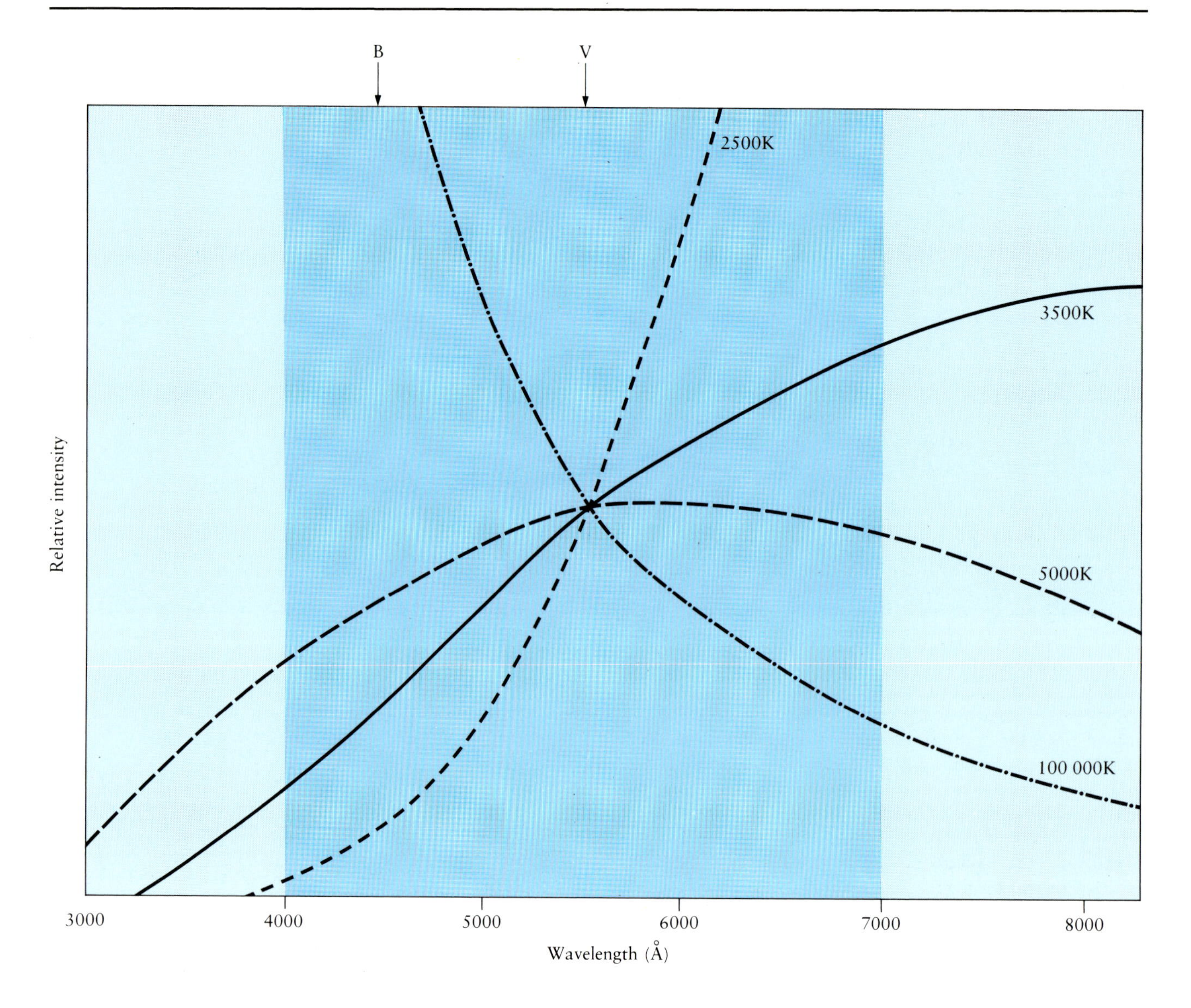

Fig. 10. *Balance of light in black-body spectra. These are spectra of black bodies of different temperatures, the sizes of the black bodies being adjusted arbitrarily so that they all radiate equally at 5500 Å. Relative to the energy radiated here, the energy radiated at 4400 Å increases with temperature from about 0.20 at 2500K to more than 2 at 100 000K. The balance of energy at 4400 Å to that at 5500 Å is thus correlated with temperature.*

calibrated in degrees Kelvin. While astronomers work at much lower light levels than most photographers, the idea of colour index is common to both. It is obviously important to both kinds of scientists to have a readily measurable index of colour, in much the same way that a doctor needs a readily measurable index of the health of a patient.

Spectral classes

Stars are not completely simple and do not have ideal black-body spectra. Overlying the surface of a star is a cooler gaseous atmosphere. Our Sun's atmosphere can be seen in its glory during a total solar eclipse (Fig. 69) when the Moon blocks out the surface of the Sun and allows us to perceive the so-called solar chromosphere and corona, which extends in tenuous form far into the Solar System, even beyond the orbit of the Earth. Atoms in the cooler atmosphere of a star absorb particular wavelengths from the continuous spectrum of the surface. Loss of these wavelengths has a measurable effect on the colour of a star, called *blanketing*, which affects the precise value of the colour index ($B-V$) at a given temperature. Different sorts of atoms can absorb light from a particular set of wavelengths. The wavelengths in this set which are actually absorbed in a given star depend on the temperature of the star's atmosphere, and this is correlated with the temperature of the star's surface. The pattern of absorption lines which appears in the continuous spectrum of the star and is therefore, like the colour index ($B-V$), a measure of the star's surface temperature.

This conclusion was not easily arrived at. The first sight of the spectral lines in the spectrum of a star was by Wollaston in 1792 who saw dark lines crossing the solar spectrum. He regarded these as boundaries between colours and missed an exciting discovery which was left to Fraunhofer who, in 1814, mapped 576 dark 'Fraunhofer lines' in the solar spectrum. He turned his spectroscope to the Moon and Venus and found identical lines in their spectra, concluding that they shone by reflecting sunlight. When he looked at the light of Sirius, Procyon and Capella he saw that their spectra too were crossed by dark lines, some of which matched the solar Fraunhofer lines, others of which did not. He concluded that the lines were caused by some absorptive power in the Sun and stars. He noted that there were coincidences between some Fraunhofer lines and the spectral lines of elements seen in the laboratory, like the two Fraunhofer lines which he called D_1 and D_2 and which are coincident with the yellow lines of sodium. The thought that the spectroscope could probe the composition of a distant star excited the scientific imagination, and Fraunhofer deserves great credit for his discovery. It was Kirchhoff who, in 1859, proved that the dark lines in the solar and stellar spectra are produced by absorption in the same vapours of the chemical elements which, if seen against a dark background, give out bright lines. Thus it was natural to regard the dark spectral lines as demonstrating the composition of the atmospheres of stars – for example, whether they contained sodium vapour and how much.

As further examples of stellar spectra were observed by the Italian astronomer Father Secchi and his British rival William Huggins in the late nineteenth century, it became clear that a progression of spectral types could be made, based on the strength of the strongest lines, identified by Balmer at about this time as coming from hydrogen. Secchi had discerned four types of stellar spectra. The first type, including those of Sirius and Vega and other white stars, was marked by strong hydrogen absorption lines (although Secchi did not initially know the element which caused them). The second group, of yellow stars like the Sun, had numerous fine lines including the hydrogen lines. The third type, of red stars like Antares, had so many lines that they shaded together in bands; and the fourth type, of deepest red stars, had carbon absorption bands, but no hydrogen lines could be discerned. By 1880 Huggins was able to give a sequence of stars which he thought represented the chemical changes in their lifetime; his list, with modern spectral types is:

Strong hydrogen lines:

White stars	**Spectral type**
Sirius & Vega	A1 & A0
α U Ma	B2
α Vir	B1
α Aql	A7
Rigel	B8
α Cyg	A2

Weak hydrogen lines:

Yellow stars	
Capella & The Sun	G8 & G2
Arcturus	K2

Banded spectra with very weak hydrogen lines:

Red stars	
Aldebaran	K5
Betelgeuse	M2

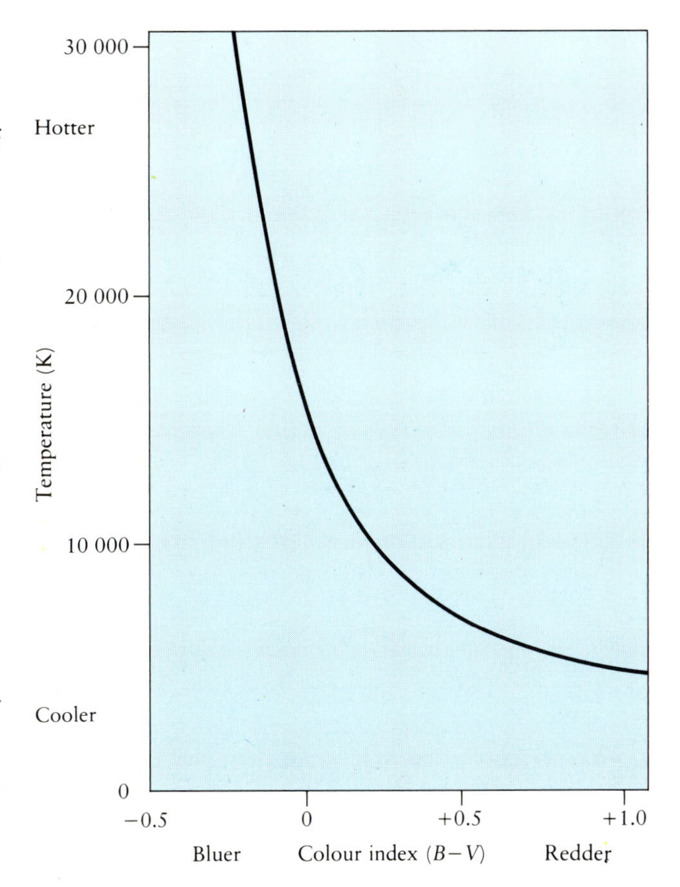

Fig. 11. *Colour index and temperature. The astronomical colour index,* B–V, *is correlated with the temperature of stars.*

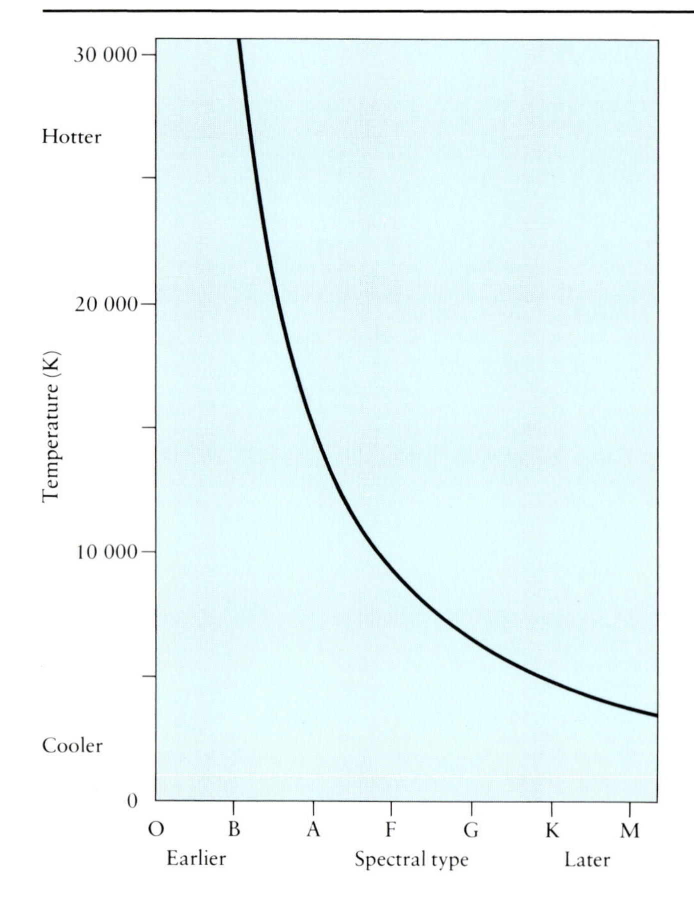

In 1886 E.C. Pickering at Harvard University began to prepare a catalogue of the spectra of all bright stars based on the Secchi classification. Secchi's four classes were subdivided into 16, denoted by the letters A to Q (omitting J). The Harvard work was published as a memorial to American astronomer Henry Draper; the first Henry Draper Catalogue, published in 1890, is on the alphabetical scheme. Notice that Huggins' list above, competitive to Secchi's, is indeed in alphabetical order of spectral type (save for the A and B grouping of the white stars between which it was difficult to distinguish).

The Henry Draper Catalogue was reworked and a new version published between 1918 and 1924. The letter classes were rearranged so that more lines besides the hydrogen series could be seen in smooth progression. Some letter classes, like C and D, proved redundant. The new order of classes was O B A F G K M and tenths of a class could be recognised (so that a B5 star appeared half-way along the B class). The hydrogen lines began weakly in O-type stars, rose to a maximum at A stars and declined in strength to M

Fig. 12. *Spectral type and temperature. The spectrum of a star is correlated with its temperature.*

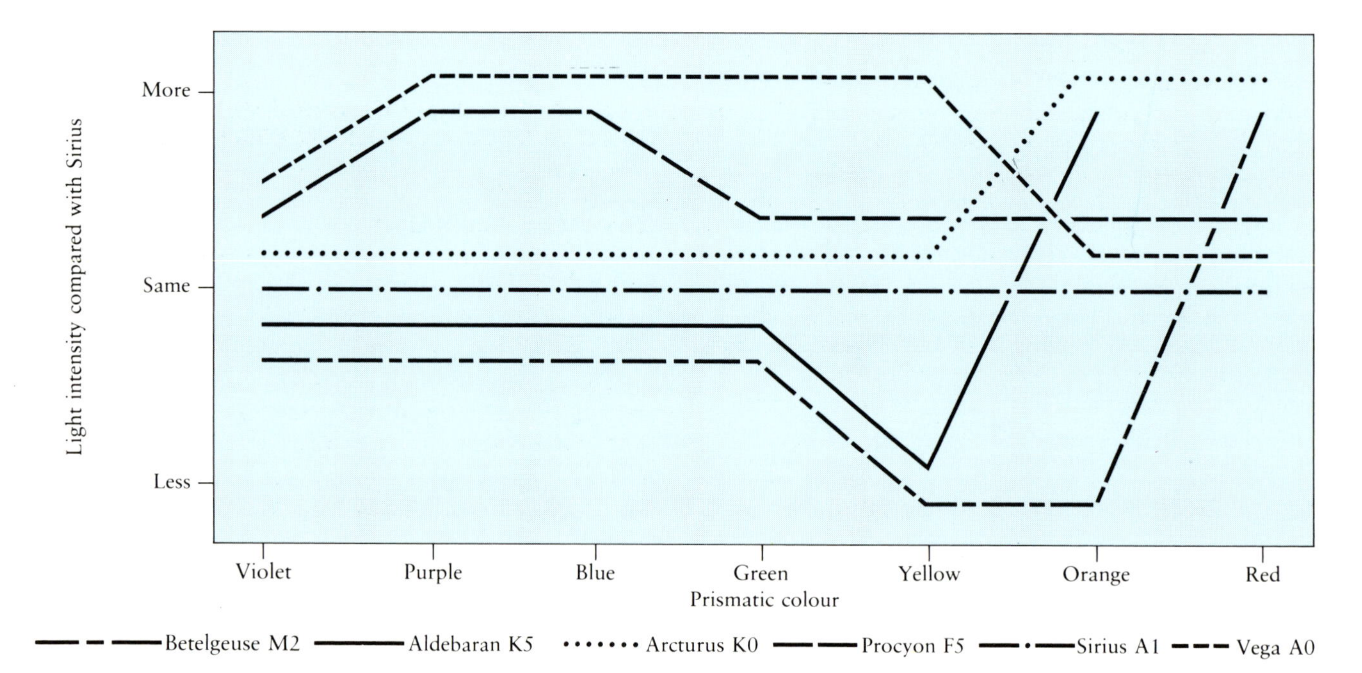

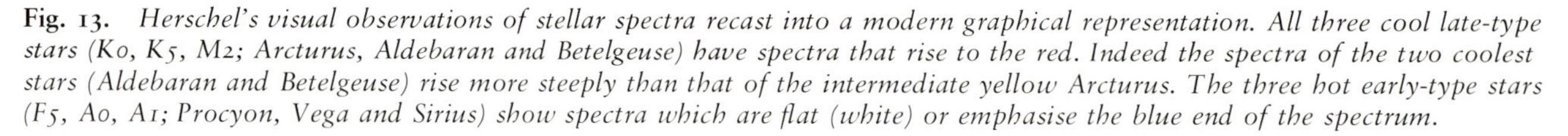

Fig. 13. *Herschel's visual observations of stellar spectra recast into a modern graphical representation. All three cool late-type stars (K0, K5, M2; Arcturus, Aldebaran and Betelgeuse) have spectra that rise to the red. Indeed the spectra of the two coolest stars (Aldebaran and Betelgeuse) rise more steeply than that of the intermediate yellow Arcturus. The three hot early-type stars (F5, A0, A1; Procyon, Vega and Sirius) show spectra which are flat (white) or emphasise the blue end of the spectrum.*

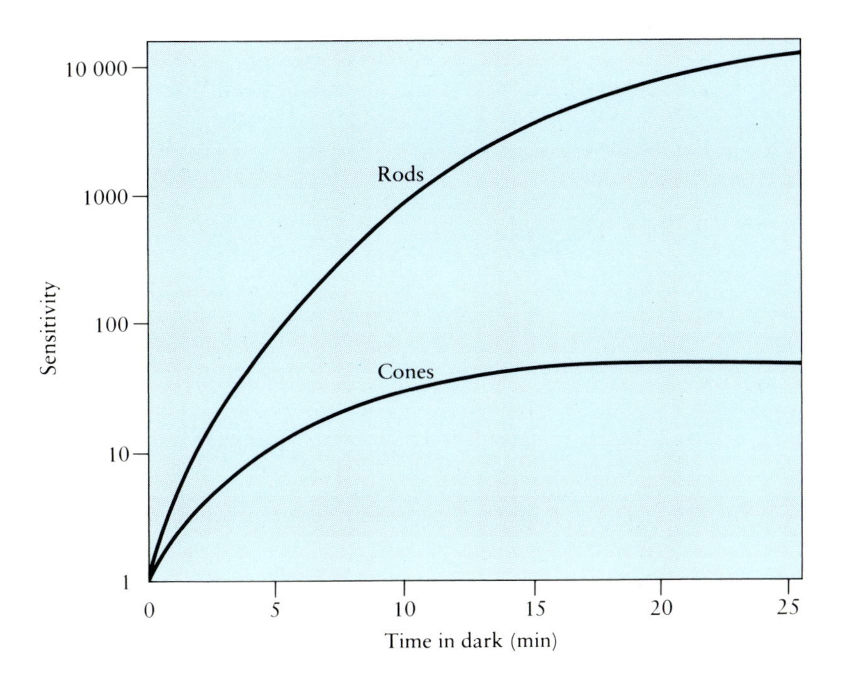

Fig. 14. *Increase in sensitivity of the eye in the dark. If the eyes are kept in darkness for some time their sensitivity increases and a given light looks brighter. The rods and cones adapt at different rates. Cones adapt fully within 5–10 min, whereas rods continue to increase in sensitivity for about a half hour, and can see much fainter lights than can cones. Rods do not perceive colour like the cones, however. Thus light from faint stars may stimulate the rods in a dark-adapted eye but not the cones.*

stars. Other spectral lines came and went as one moved down the sequence of spectral types. The dual evidence of star colour and behaviour of spectral lines made the Harvard astronomers believe that the spectral sequence indicated changing physical properties, not composition. They proposed that the O B A F G K M sequence was in order of temperature from hot (O-stars) to cool (M-stars). They formed the sequence empirically, and it was not until the changing strength of spectral lines was explained theoretically by M.N. Saha in 1920–21 that it became accepted that the sequence of spectral types was indeed a temperature sequence. The relation between spectral type and temperature is given in Fig. 12.

Star colours and spectra

We can actually relate spectral types of stars to directly perceived shapes of their spectra (Fig. 4), by using remarkable observations made by Sir William Herschel in 1798. He was following suggestions by Thomas Collinson and Sir William Watson in 1783 that he should apply spectroscopy to the study of starlight. He cannot be said to have followed the suggestions with alacrity and M.A. Hoskin (1963) has noted that Herschel took little interest in making observations which he had no hope of interpreting. Nevertheless, 'by a prism applied to the eye-glasses of my reflectors', Herschel inspected the light of a few stars of the first magnitude (Table 2). By referring to Sirius as a standard we can plot his observations as spectra (Fig. 13). They generally confirm the theoretical links between spectral types, the continuum shapes and star colours.

TABLE 2. SIR WILLIAM HERSCHEL'S OBSERVATIONS OF STAR SPECTRA (1798)

Observation (quoted from Herschel, 1814)	Spectral type
The light of Sirius consists of red, orange, yellow, green, blue, purple and violet	A1
Alpha Orionis (Betelgeuse) contains the same colours but the red is more intense, and the orange and yellow are less copious in proportion than they are in Sirius	M2
Procyon contains all the colours but proportionally more blue and purple than Sirius	F5
Arcturus contains more red and orange and less yellow in proportion than Sirius	K0
Aldebaran contains much orange and very little yellow	K5
Alpha Lyrae (Vega) contains much yellow, green, blue and purple	A0

Fig. 15. *Star trails. Stars have left their coloured trails over the dome of the Anglo-Australian Observatory in this 10.5-hour time exposure. The stars circle the South Celestial Pole above the dome.*

Seeing colours

The spectra of the stars and astronomical theory show that stars ought to look coloured, but even skilled observers of star colours see nothing dramatic, although the colours are there.

The major reason is physiological. The human eye consists of a lens which can focus light on to a sensitive screen, the retina. The retina contains two kinds of light receptors called *rods* and *cones* because of their appearance under the microscope. The rods and cones are intermingled throughout the retina. However, there is a greater proportion of rods at the edges of the retina while the cones are concentrated toward the centre of the retina. The cones perceive colour whereas the rods give vision only in shades of grey. The cones function only when activated by light of sufficient intensity. The rods are more sensitive than the cones to dim light (Fig. 14). Starlight seen with the unaided eye activates the rods but does not activate the cones to the same degree. Thus the colour of the stars is cloaked by a physiological deficiency.

The eye's deficiencies are demonstrated by the following photographic experiment. If you can go to a place away from city lights and can guarantee a night free from clouds, rain, dew or camera thieves, you can take a picture to show star colours easily. Load your camera with a fast colour film, put it on a firm support and open the lens to a large diaphragm setting (use a setting between f/2.8 and f/4 with a fast film of speed between 200 and 400 ASA). Focus the lens on infinity, open the shutter and keep it open as long as you can. You'll get a picture like Fig. 15.

The point about this picture is that the star trails are distinctly coloured with a range of hues from pale blue to deep red. Only the colours of the brightest stars in this part of the sky are visible to the naked-eye observer.

If the astronomer has access to a telescope the situation is a little improved. Of course there are many more stars to be seen with such an instrument. Still only the brighter stars seem weakly coloured. The colour can be made more apparent when the telescope is defocused and the star images are no longer point-like. Under these conditions the brighter stars are seen as discs of light and may have distinct shades of blue, yellow and orange. The most extreme example of this phenomenon is seen when, with no eyepiece, you place your eye near the focal plane of the telescope so that all the light from a single bright star enters its pupil. What is seen is the mirror or lens of the telescope filled with starlight, a condition named *Maxwell's view* after the Scottish physicist who used it in his colour mixing experiments. With a selection of suitably bright stars quite subtle changes in hue can be seen and it becomes evident that some stars have quite distinct colours.

Why is it that it is easier to see colour in a defocused star? The same effect can be seen from the window seat in a high-flying aeroplane during a night flight. As the towns and cities pass below, the distribution of various kinds of street lighting can be seen; suburbs may be lit with yellow sodium vapour lamps while bluish mercury tubes may illuminate the highways. From an airliner at 30 000 feet these two types are easily distinguished by their colours, but it is the colour of the illuminated area beneath each lamp which gives the strongest indication of colour, not the much brighter lamp itself, effectively a star-like point when seen from this altitude.

So the perception of colour is not only a matter of activating the colour-sensitive cones in the eye; there are complex psychological effects too. The examples above demonstrate one such effect: the eye is relatively insensitive to the colour of small sources. The effect has the technical name (Weitzman and Kinney, 1969) of *small field tritanopia* (tritanopia means colour-blindness of the third kind). Scientific studies have shown that as the perceived size of coloured patches is decreased, the colours seem to disappear in pairs. The colours of yellow and blue disappear first, followed by green and red.

Another curious fact emerges from our high-flying observations. We are well aware now that at low light levels colour discrimination becomes difficult and that at very low levels colour sensitivity ceases altogether. Directly beneath a yellow sodium street lamp the light reflected from the road is bright enough to appear coloured. As the illumination level falls off away from the lamp, from our high vantage point we would expect to see a yellow pool of light surrounded by a grey halo, colourless because the light level is too low to activate the cones of the retina. The circumstantial evidence is too strong, however, and our brains assume that the yellow colour extends to the visual limit of the halo. We believe what we think we see.

Even though a colour may be difficult to see individually, it may show up as more intensely coloured when seen together with a contrasting colour. The group of painters known as the Fauves, led by Henri Matisse, used this fact in the assault on the senses which gave the group its nickname of 'wild beasts'. Thus the colour of stars can be more easily perceived when stars of contrasting colours are seen together. Particularly striking pairs of stars which give contrasting impressions even with telescopes of modest size are listed in Table 3 (see also Fig. 16). Rather surprisingly, globular clusters show strong contrast amongst their stars. Normally the superficial impression is that such clusters (Fig. 17) are homogeneous, with stars of only one kind, and one colour. It is true that the large majority of the brighter stars are of one kind, but contrasting faint bluer stars may be picked out from the brighter, whiter-seeming

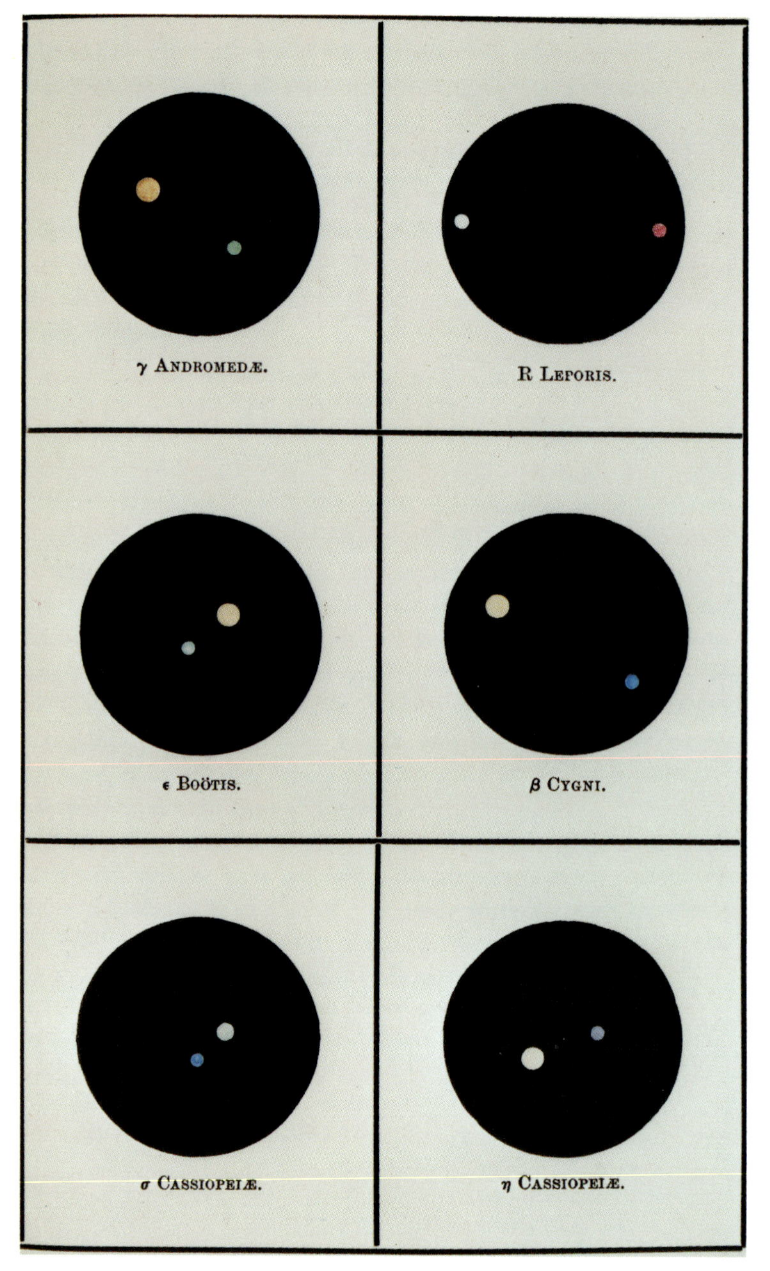

Fig. 16. *Contrasting double stars. Six pairs of coloured stars are illustrated in this hand-tinted reproduction from Chambers'* The starry heavens *of 1877.*

TABLE 3. CONTRASTING DOUBLE STARS (CHAMBERS, 1877)

Name	Brighter star	Fainter star
η Cas	Yellow	Purple
α Psc	Pale green	Blue
γ And	Orange	Sea green
ι Cnc	Orange	Blue
ε Boo	Pale orange	Sea green
ζ CrB	White	Light purple
α Her	Orange	Emerald green
β Cyg	Yellow	Sapphire blue
α Cas	Greenish	Bright blue

stars in the illustration. In some brighter globular clusters the few so-called blue horizontal-branch stars stand out so clearly from the mass that they can easily be picked out by eye. The colour differences are quite subtle but are readily seen by contrast.

Perceiving colour in stars by eye is even more complex because the observations are necessarily made under dark conditions. At low light levels, the pupil of the eye opens fully to admit more light. As any photographer will tell you, lenses produce poorer images at full aperture than they do with the iris

Fig. 17. *Hodge 11. The brighter white stars in this globular cluster in the Large Magellanic Cloud (LMC) contrast with the fainter bluer ones. The subtle colour difference is enhanced by contrast of the stars in such proximity one to another. Normally globular clusters are thought of as containing old red stars; see p. 87. Globular clusters in the LMC are bluer than ones in our own Galaxy, leading to interesting speculations on their relative youth, and why the LMC should be different from our Galaxy in this respect. Hodge 11 is 'unusual in the presence of a number of very blue stars' (Gascoigne, 1966). For the first time this photograph shows how they concentrate to the centre of the cluster.*

diaphragm partly closed. The image formed by the dark-adapted eye is therefore much inferior to that projected on to the retina under bright light. In particular an effect known as *chromatic aberration* will be present, producing images with coloured haloes. The remarkable image-processing capacity of the brain is able to reconstitute these rather large images with their coloured haloes into point sources but in doing so colour information tends to be lost.

Opening its pupil is not the only way that the eye adapts to darkness: the rod and cone receptor cells become more sensitive. They adapt at different rates (Fig. 14). The cones become about 40 times more sensitive after about 7 minutes but do not improve much further. The adaption to darkness of the rods continues for more than half an hour; they become several hundred times more sensitive to light than the cones. Because the rods are more sensitive to blue light, dim red light becomes more difficult to see than dim blue light with a dark-adapted eye (Fig. 18). This is known as the Purkinje effect and means that, not only is the colour of dim light difficult to perceive at all, but that any perception of the colour is liable to be mistakenly too blue. The predominance of blue and green among the colours of the fainter stars of Table 3 is partly due to this. A further use of the colour separation between rods and cones can be seen in astronomers' use of red torches at night. Red light is seen by the cones, but does not affect the rods, so use of a red torch to inspect a chart of the stars leaves the dark adaptation of the rods set at a level still useful to see the stars through a telescope – we say that red light is less dazzling.

Colour discrimination at the very threshold of colour sensitivity is in fact unpredictable. In an experiment by Bowman (1961) dark-adapted observers were shown flashes of yellow light covering a field of 2°, about four times the diameter of the Full Moon. The intensity of the light was gradually increased and the subjects of the experiments were asked to describe what they saw. Initially they reported simple perception of a flash without naming colour. As the brightness was increased above the just-detected level, the observers began to report that some of the flashes appeared red, some green and some white. Only at a considerably higher level of illumination was yellow reported along with the (non-existent!) red, green and white lights. Quite clearly then, the eye is poor at defining the colour of any kind of source at low light levels (Hunt, 1952).

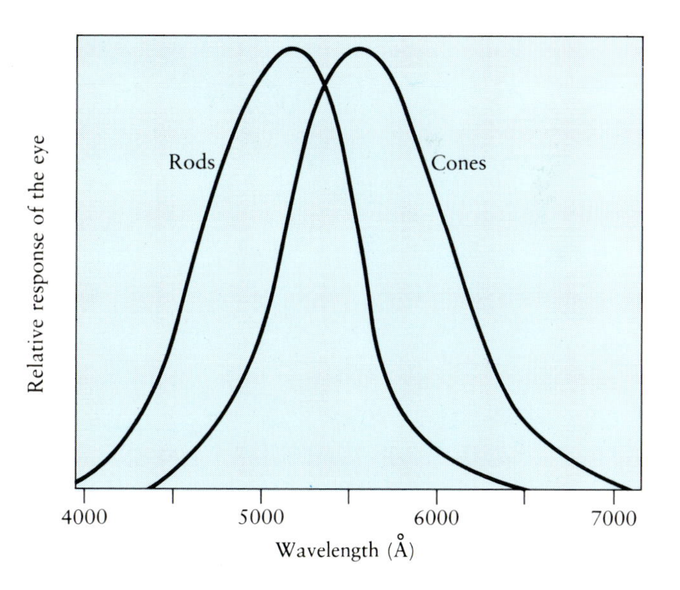

Fig. 18. *Overall spectral sensitivity of the eye. The two curves represent the brightness-producing capacity of light of different wavelengths. On the right is the normal bright-light response of the eye, with peak sensitivity near 5500 Å (yellow-green). Light of wavelength 7000 A only just stimulates the cones in the retina. Note how insensitive the eye is to blue light (4000–5000 Å). The left-hand curve shows the sensitivity of the dark-adapted eye. Blue sensitivity is increased, red sensitivity decreased.*

The colour of moonlight

As an example of the difficulty in seeing colour consider the same scene illuminated by day and by night. The difference in colour is well known, and was described in Ursula Vaughan Williams' poem for her husband's cantata *Sons of light* (Lyrita records SRCS 125):

> *Rise moon, cold crescent of reflected fire,*
> *make night all silver now gold day is done.*

By the light of the Sun we see the colours which exist in the scene, but by moonlight the colours are imperceptible. The cinematic convention for moonlit scenes is to photograph them through a blue filter to suppress colour and simulate the Purkinje effect (this cinematic technique is called *nuit americain* by French film critics). Moonlight is only reflected sunlight, but of lower intensity. The colours are there and the same but we cannot see true colour in moonlight because the colour-sensing parts of our eyes are not activated adequately by low-intensity light.

None of this must be taken to mean that the eye is a poor detector of light – just of its colour. Photons are individual particles of light, the smallest quantities in which light exists. Detection of individual photons is regarded as the ultimate in performance of a light-sensitive device. Under the right conditions the arrival of a very few photons on the retina is sufficient to produce a response (Hecht, Shlaer and Pineune, 1942), a performance only recently bettered by electronic sensors. For instance, the light of the Full Moon is about half a million times fainter than noon sunlight but is quite sufficient for a dark-adapted man to find his way about with little difficulty. In fact a fully dark-adapted person is quite capable of finding his way around in unfamiliar terrain by the light of the stars alone. He might move with great caution but he is not completely sightless under these conditions. It helps if the Milky Way is overhead or if Venus is high in the sky. The eye is sufficiently sensitive even to see the shadow cast by Venus. Laurens van der Post (1963) writes in his autobiographical novel *The seed and the sower*:

. . . Celliers drew my attention to the evening star which was hanging so large and bright in the faded track of the sun that it looked as if it might fall from sheer weight out of the shuddering sky.

'It's odd,' Celliers was saying, 'How that star seems to follow me around. You should have seen it as I saw it in a winter sky over the high veld of Africa or one night over the hills before Bethlehem. I spotted it even in full daylight from the jungles of Bantam – but I've never seen it more lovely than now. There's light enough in it to-night to fill both one's hands to overflowing.' He broke off to peer down towards the ground at our feet where the earth was still wet and glistening from a heavy downpour in the afternoon. 'Look!' he exclaimed, his voice young with astonishment. 'It throws a shadow as well. Look just behind you in the wet. There's your star shadow following you faithfully around. How strange that even a star should have a shadow.'

This ability to see faint light and to adapt to varying light is all the more remarkable when we compare the eye to a camera. If we set a camera to take a well-exposed photograph of a scene in bright sunlight with an exposure time of one five-hundredth of a second (0.002 s), this same system would need over an hour to record the same scene by the light of the Full Moon. In practice much longer is required due to the failure of the film to record faint light as efficiently as it does bright light. Even so, it is, however, possible to take colour pictures by moonlight which show quite

Fig. 19. *These two pictures of the Anglo-Australian Observatory dome were taken 13 h apart, one by sunlight (left), the other by the light of the Full Moon (below). Both were made on the same roll of 400 ASA colour film, one with an exposure of 0.002 s at f/22, the other, 11 min at f/4, so that the difference in exposure factor is 10 million. The moonlight scene differs from the daylight image in showing star trails (in a blue sky), the Large Magellanic Cloud (to the right of the top of the dome), and the blurring of the gum trees as they move in the wind. Although the colours of the moonlit scene are revealed by long-exposure photography, moonlight does not activate the colour-sensing cones in the eye's retina.*

clearly that the full range of colours is still present as in Fig. 19. The colours are there in both cases: in the dimmer moonlit scene they were not perceived. The photograph reveals them because both photography and electronic sensors win over the eye in the long run by being able to sum up or integrate incoming light over very long periods, sometimes many hours. The eye on the other hand has an integration time of much less than 1 s, so, while it can detect very few photons, it cannot assemble many successive and infrequent arrivals into this full image.

To summarise, in visual astronomy the eye is presented with the worst possible situation for perceiving colour. It is faced with bright, point sources superimposed on a very dark background. The pupil of the eye is wide open, which results in poor optical images being formed on the light-sensitive retina. Though bright in relation to the background, the point sources are near the threshold of detection for colour vision.

There are therefore physiological and astronomical reasons which hinder any attempt to determine star colours. Nonetheless, star colours are not an astronomical myth (Cohen and Oliver, 1981) and the pale, tinted sequence from blue to red can be perceived by the careful observer.

Colours of stars

To see how accurately the colours of bright enough stars can be perceived we have investigated the observations of star colours by nineteenth-century astronomers who systematically observed double stars but were unprejudiced by knowledge that their colour meant temperature. The first was F.G.W. Struve (1827) who observed double-star orbits with various refracting telescopes at the observatory in Dorpat in Estonia and noted the stars' colours. The second was Admiral W.H. Smyth (1844, 1864), author of the famous *Cycle of celestial objects*, a long list of double stars, nebulae and clusters for astronomical sightseeing. Smyth was very interested in star colours and wrote a monograph called *Sidereal chromatics or colours of double-stars* in which he discussed his observations and those of others, including his son Piazzi Smyth, whom the Admiral had inveigled into observing star colours during a site-testing expedition to Tenerife. Of the two astronomers it was Struve, the professional astronomer, who was the more conservative. He classified stars into only ten basic colours which he expressed in a twin-branched gradation (Table 4). Occasionally, this simple scheme defeated his eye and Struve had to use more complex colours like *olivaceasubrubicunda* (pinkish-olive)!

Struve's work was translated into English and edited by T. Lewis (1906) for the Royal Astronomical Society. There is in the library of the Royal Greenwich Observatory a copy of the memoir in which an unknown astronomer, with an inky nib, has stylishly scrawled the spectral type of most of the stars Struve observed. The spectral type obviously refers to the brighter of the star pair and it was easy to correlate the colour of this star with the spectral type. Because we now know that colour and spectral type are correlated with temperature (p. 8), it is appropriate to order the colours into a sequence reminiscent of those

TABLE 4. STRUVE'S STAR COLOURS

– *egregie albae*
very white
– *albae*
white

– *albaesubflavae* yellowish white	– *albaesubcaeruleae* bluish white
– *subflavae* yellowish	– *subcaeruleae* bluish
– *flavae* yellow	– *caeruleae* blue
– *aureae* golden	
– *rubrae* red	

To this Struve added stars of mixed colours:

– *albae subvirides* greenish white	– *cinereae* ashy
– *subvirides* greenish	– *purpureae* purple
– *virides* green	

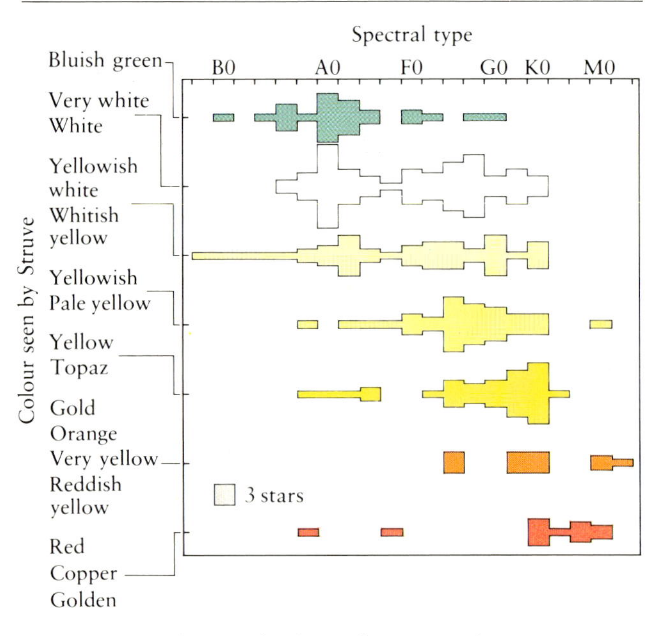

Fig. 20. *Correlation of colour of stars seen by F.G.W. Struve against HD spectral type. The yellow, yellowish and white stars are so many that only a sample is plotted. There is a clear trend of spectral type with colour.*

which a hot body would show as it cooled – blue-hot, white-hot, yellow-hot and red-hot is the natural progression of the colours and the temperature, and O B A F G K M is the progression of the spectral types. To minimise the psychological and physiological problems in colour perception, to which we devoted a previous section, we looked only at Struve's observations of bright stars, in the expectation that faint ones would show confusing results.

Most stars are called *white* by Struve. There are also large numbers of *yellowish* and *yellow* stars, and it was adequate to sample the spectral types of these stars. Some of the colour classes could be amalgamated (all the *greens* and *blues* for instance) because there were few stars of each colour and they could not be separated significantly. This left seven subdivisions (Fig. 20). Although the *green and blue* amalgamated class is one of the most tightly defined groups, the majority being A, the *white* shade contains a similar concentration of the same A stars. Struve thus found it difficult to distinguish blue from white stars. The *whitish* and *yellow* shades refer to stars from A to K in spectral type, but there is a clear progression in average spectral type, which refers to cooler stars, as the colour gets yellower. The spectral types of stars labelled with orange shades range from F to M type, and the stars called *red*, *copper* or *golden* are almost all K and M types.

There is an unknown contribution to the scatter in the correlation from the spectral types and from amalgamation of Struve's colours, but the evidence is that Struve could distinguish with some reliability several star colours.

How did Struve compare with Admiral Smyth? Smyth (1864) classified stars into a large number of shades. He must have become self-conscious after some contemporary expressions of doubt that he could distinguish so many hues. In his *Colours of double-stars* he lists some 'inexact epithets' which he admits he had used in the past and gives what he felt he probably meant (Table 5). His career as an armchair sailor is revealed by the surprising knowledge of the colours of semi-precious minerals, fruits and flowers. The table of colours contains some peculiarities – vanilla, to moderns, probably means the colour of white ice-cream not the chocolate colour of the bean – and one must admire the subtlety of 'bright but pale yellow'.

Fig. 21 shows the correlation of the colour of the bright stars in the first half of Smyth's *Cycle of celestial objects* with $B-V$ colour index given in the *Bright star catalogue* (Hoffleit, 1966). As explained on p. 7, colour index, like spectral type, is a measure of stellar temperatures and should correlate well with colour, and again we avoid the problems of colour perception in faint stars. It was necessary to amalgamate Smyth's colours more ruthlessly than Struve's, but this still left eight subdivisions. When plotted against colour index, the subdivisions in colour match well with Struve's, and Fig. 21 shows common features with Fig. 20. Smyth succeeded in distinguishing the same basic half dozen star colours as Struve but there is the same difficulty in grading the *blue-white* and *red* ends of the spectrum, and a comparatively fast switch in yellow stars.

The astronomical reasons for this have already been implied on p. 6. The spectrum of a hot star peaks in the ultraviolet and is relatively flat in the visible region, with a gentle decline to the red. The uniform spectrum of a hot star seems white, perhaps with a slightly blue look, at most as blue as the sky (see Fig. 107). If we now imagine the star cooling and becoming redder, the peak of the spectrum passes through the visible region of the spectrum to the red.

TABLE 5. SMYTH'S STAR COLOURS

Inexact epithet	What was probably meant
Amethyst	Purple
Apple green	Brownish green
Ash colour	Pale dull grey
Beet hue	Crimson
Cinerous	Wood-ash tint
Cherry colour	Pale red
Cobalt	Bluish white
Creamy	Pale white
Crocus	Deep yellow
Damson	Dark purple
Dusky	Brownish hue
Emerald	Lucid green
Fawn-coloured	Whitey-brown
Flushed	Reddened
Garnet	Red of various shades
Golden hue	Bright yellow
Grape red	A variety of purple
Jacinth	Pellucid orange tint
Lemon-coloured	Bright but pale yellow
Lilac	Light purple
Livid	Lead colour
Melon tint	Greenish yellow
Orpiment	Bright yellow
Pale	Deficient in hue
Pearl colour	Shining white
Plum colour	Pale purple
Radish tint	Dull purple
Rose tint	Flushed crimson
Ruby colour	Pellucid red
Ruddy	Flesh-coloured
Sapphire	Blue tint
Sardonyx	Reddish yellow
Sea green	Faint cold green
Silvery	Mild white lustre
Smalt	Fine deep blue
Topaz	Lucid yellow
Vanilla tint	Dark brown or chocolate

Because the black-body curve drops sharply to the short-wavelength side of the peak, the blue end of the spectrum is quickly suppressed and the colour switches rapidly to a red colour. As the peak moves into the infrared we see the tail of the black-body curve and the star colour gets redder still.

Thus the shapes of the correlations in Figs 20 and 21 are immediately understandable for astronomical reasons. There is a further physiological reason for the difficulty of distinguishing blue from white stars. Fig. 18 shows how insensitive the eyes are to blue light. Even under dark-adapted conditions, dark adaption to perceive blue is thus less effective than to perceive red, and blue stars may appear colourless although red stars of the same brightness appear red.

We have here the two reasons why the nineteenth-century visual observers found red stars so fascinating and startling. Added to these is the fact that the red stars are often very bright, supergiant stars, which represent a brief stage in the lifetime of stars (see Chapter 4). They are thus prominent but rare, and this makes them all the more noteworthy, like the triangular postage stamps of the Cape of Good Hope (1853–1864). Several great visual observers 'collected' red stars during their long hours sweeping the night sky. Building on the casual discoveries and short lists by Herschel, Lalande and De Zach, the Danish astronomer Schjellerup made a comprehensive list of red stars in 1874, over 200 of which appear in a revized catalogue by Chambers (1877). The Rev. T. E. H. Espin made a fine catalogue of red stars; the brighter members of these lists appear on the celestial maps of Norton's *Star Atlas* (e.g. Fig. 37*b*), identified with the letters E–B (for the Espin–Birmingham catalogue), R (for *red*) and Ru (for *ruddy*). Identifying them with field glasses on a moonless, clear, warm summer's night can pass a pleasant couple of hours.

The colour index and spectral type of stars of a given colour assigned independently by Struve and Smyth agree well, and both observers therefore have about the same colour sensitivity. This agreement on colour is not universal amongst all astronomers. Struve (1827) gives lists of those bright double stars of which both he and Sir William Herschel had observed the colour. It is difficult to compare the colours directly because the two men do not use standard names. Nevertheless, the above comparisons between colour and spectral type and between colour and colour index suggest the following six basic possible star colours, of which not all are equally distinguishable: in order of decreasing temperature they are *blue or green, white, off-white, yellow, orange, red*. We have assigned Herschel's and Struve's colours to one of these classes and plotted a comparison (Fig. 22). The stars which Herschel saw as blue or green were almost without exception seen to be the same by Struve, but Herschel was much more inclined to see a redder colour than Struve. He saw reddish tints in white stars (*off-white*) where Struve saw none (*white*). The stars Herschel classified as *red* Struve could see as any shade from *blue* to *red*, but most often *yellow*.

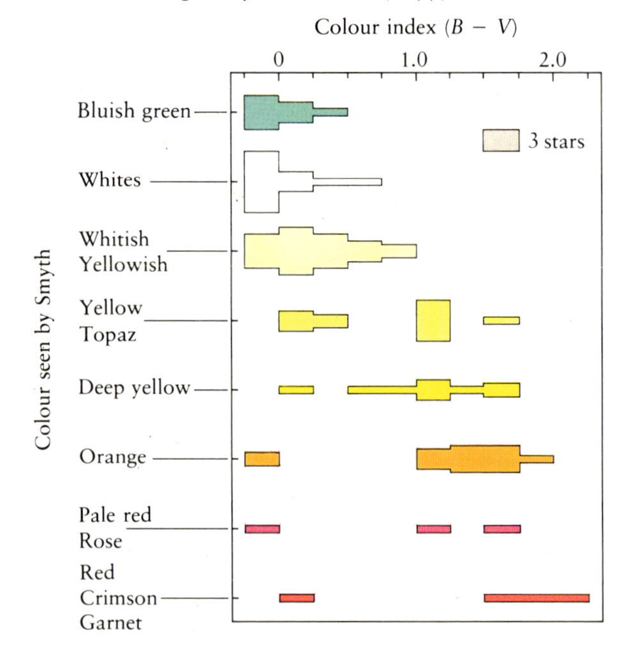

Fig. 21. *Correlation of colour of stars seen by W.H. Smyth and photoelectrically measured colour index (B−V). Smyth found it difficult to distinguish whites and blues but there is a trend amongst the yellow stars which persists less definitely through the orange and red stars.*

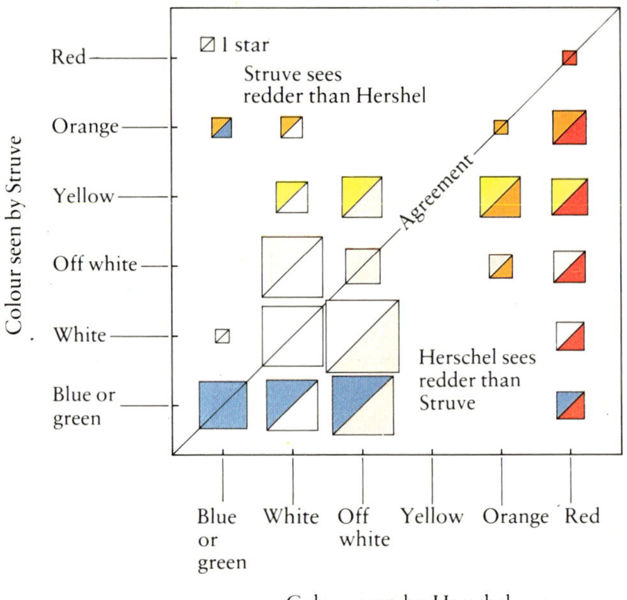

Fig. 22. *Comparison of Struve's colour estimates with those of Herschel. Most points lie in the lower right half, signifying that Herschel emphasised the red in perceiving star colours.*

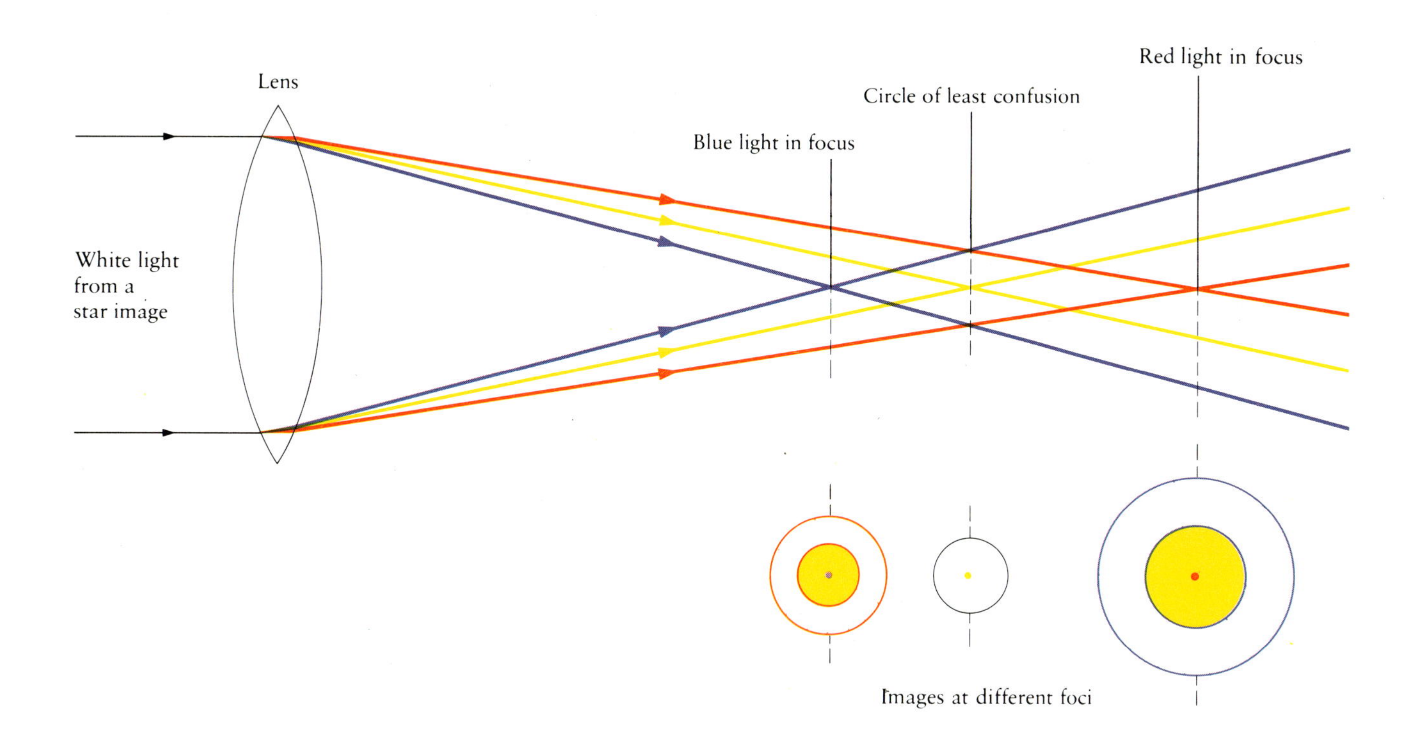

Instrumental effects?

In his double-star observations, Struve used a $9\frac{1}{2}$-inch refracting telescope. It was the first truly equatorially mounted telescope, in which one axis pointed to the north celestial pole, enabling stars to be tracked by motion around this axis alone. The lens was made by Fraunhofer himself, the whole telescope being put together with the utmost workmanship. The light-gathering lens was made of two components of crown glass and flint. Fraunhofer had used his discovery of the spectral lines in the solar spectrum to determine the refractive indices of different kinds of glass at different colours, and this enabled him to match lenses of two kinds to correct for the coloured haloes (Fig. 23) which single lenses produce around stars (chromatic aberration). The reason for this aberration is that the refractive index of glass produces a stronger bending of blue light than of red, so that the image of a white star appears as a succession of concentric blue, yellow and red images. Depending on how the eyepiece which views this succession is focused, the white light image has a coloured appearance.

The principle for making an achromatic lens had been discovered 50 years before Fraunhofer constructed the Dorpat refractor, by John Dollond, following work by Euler and Hall. Dollond's techniques were improved by Fraunhofer using methods which he took with him, as trade secrets, to his grave. Modern investigations show that Fraunhofer

Fig. 23. *Chromatic aberration. Simple lenses do not refract all colours equally but focus blue light nearer the lens than red light. The best focus of parallel starlight is the spot where the blue and red images are of equal diameter. Inside focus, blue point images have red haloes; outside focus, red images have blue haloes.*

slightly overcorrected the chromatic aberration in the lenses which he made, with the result that a blue halo was left surrounding stars. It is sometimes said that this influenced Struve's estimates of the colour of the stars which he saw with the telescope, and that he tended to estimate them too blue. Smyth also used refractors, particularly a 5.9-inch refractor made by Charles Tully for Sir James South. Smyth enthused about its performance:

> *On repeated trials, I find the instrument bears its highest magnifiers with remarkable distinctness, as is especially evinced by the roundness of small discs, the dark increase of vacancy between close double stars . . . I have therefore reason to presume that the curves of the lenses are in exact chromatic and spherical aberration throughout.*

William Herschel's enthusiasm for the reflecting telescope which he made and used is well known. Indeed it was his failure to correct chromatic aberration in short-focal-length lenses as well as the impossibility of obtaining large enough pieces of transmitting glass that spurred him into making reflectors. Mirrors have no chromatic aberration. This and the known overcorrection of Struve's refractor have led to the suggestion that instrumental reasons explain the difference between Herschel's and Struve's observations of colours. The agreement of Struve and Smyth on the other hand points to Herschel as the odd man out, and suggests that the difference is a psychological or linguistic one. Whatever the reason for the relative differences between these three great observers, whether subjective in their individual perception of colour, or objective in their instruments' chromatic correction, the comparisons between Struve's observations of colour and spectral type and between Smyth's observations and colour index both confirm the reality of the colours of stars.

Spectral typing by seeing colour

Colour index and spectral type are related to colour-temperature. How accurately then can the eye measure temperature? Struve's and Smyth's observations show a lot of scatter; but they were selecting colours from the very wide range found in daily life, whereas in stars few colours are possible (those given essentially by black-body curves). If the eye could be educated to select only from the shades which are reasonable, how well might it do? Observations which test this have been published by M. Minnaert (1954). He gives his own scale of star colours based on a body cooling from very hot to cold – the same progression as the sequence of spectral types and colour index (Table 6). Minnaert also gives some examples of colours of bright stars and planets observed with the naked eye by some friends of his. No stars (even Sirius and Vega) are listed as blue or bluish, and there were apparently no red stars worthy of note and visible to the naked eye at the time of the observations. Even Antares, outstanding to the eye for its red colour compared with the other stars of Scorpius in its vicinity (Fig. 24, see also Fig. 37), is listed as orange, not red. The shades of yellow – white yellow, light yellow, pure yellow and deep yellow – were subtly distinguished.

How well does Minnaert's colour scale stand up against objective measurements of colour? We can find out by plotting (Fig. 25) colour scale against colour index ($B-V$), from the data in Table 6. The group of white stars is not really distinguished from the white yellow stars, because the $B-V$ colours of the two groups do not show a real difference. The shades of yellow on the other hand are well separated, with no overlap between groups in the spread of $B-V$, and scarcely any overlap into the orange. The general correlation between colour scale and colour index is excellent!

In fact, the scatter of points in Fig. 25 about

Fig. 24. *Antares and the stars nearby. Antares is the bright orange star at the bottom of this picture. Other stars in the field range in colour from white to red.*

the best straight line fit is only about a quarter of a magnitude in $B-V$, so the error in colour scale measurement is one-eighth of the range from the bluest star observed to the reddest. We can thus say that Minnaert can distinguish eight colours in stars. Notice that the range of spectral types of the stars which he observed (B type to M type) cover six spectral classes. By looking at its colour Minnaert could accurately spectral type a star to within one letter class!

The conclusion that it is possible to do this has been confirmed by the reminiscence of Leslie Morrison of the Royal Greenwich Observatory, who is an astronomer observing the positions of stars with a transit circle telescope. This is one of the few professionally made observations in astronomy still executed with the naked eye (although a photoelectric device has been made for the transit circle Morrison is now building!). Morrison says that as he watched a star approach the cross-hair in the telescope for a position measurement to be made he used to challenge himself to guess the star's spectral type by its colour, and could consistently guess the correct catalogue value. Thus, experienced naked-eye observers are able to sort stars into one of the seven letter classes O B A F G K M fairly accurately by eye.

The colours of the Sun and Moon

Minnaert's colour scale gives us a way to estimate the colour of the Sun, if we could view it as a bright star. It has a spectral type of G2 and a colour index of $B-V=0.62$, and is almost a twin of the first magnitude star Alpha Centauri, one of the pointers which identify the Southern Cross and the South Celestial Pole. The Sun viewed from Alpha Centauri looks like Alpha Centauri viewed from here. The colour scale value for $B-V=0.62$ is 3.6, or light yellow, which is how Alpha Centauri appears. Struve saw G3 stars as a pale yellow (Fig. 20) and Smyth's observations are consistent with this. All the indications are that the Sun would be a pale yellow colour if we could bear to look at its dazzling face in clear sky, and the orange represented in Fig. 7 is too deep a colour. It is rather strange that when yellow sunlight falls on clean paper we agree to call the paper 'white', as we do snow and clouds. Our perception of colour for brightly illuminated extended objects is different from that for stars. However, the Full Moon, which when it reflects sunlight reddens it slightly (to $B-V=0.91$), may look a pale yellow colour when we

TABLE 6. MINNAERT'S COLOUR SCALE

Colour scale	Colour	Examples	Colour scale value	Colour index ($B-V$)
−2	Blue	–		
−1	Bluish white	–		
0	White	α CMa (Sirius)	0.8	−0.01
		α Lyr (Vega)	0.8	0.00
1	Yellowish white	–		
2	White yellow	α Leo (Regulus)	2.1	−0.11
		β UMa (Merak)	2.3	−0.02
		α CMi (Procyon)	2.4	0.41
		α Aql (Altair)	2.6	0.22
3	Light yellow	Venus	3.5	0.79
		Jupiter	3.6	0.6
		α UMi (Polaris)	3.8	0.6
4	Pure yellow	α Boo (Arcturus)	4.5	1.24
		Saturn	4.8	0.9
		α UMa (Dubhe)	4.9	1.06
5	Deep yellow	β UMi (Kochab)	5.8	1.49
6	Orange yellow	–		
7	Orange	α Sco (Antares)	7.5	1.83
		Mars	7.6	1.41
8	Yellowish red	–		
9	Red	–		

see it high in the sky. Its small angular size and situation in the dark sky evidently modify our perception of its colour.

The physiological responses of our eyes and brains to the astronomical facts of the colours of the Sun and the Moon have, during the evolution of the human race, become confused with our psychological responses to the circumstances under which we have habitually perceived the colours. The yellow of sunlight has become associated with the heat of the Sun and we are conditioned to think of sunlight's golden colour as 'warm'. The light of the Moon, on the other hand, demonstrated scientifically to be if anything yellower than the Sun, we perceive as blue like *nuit americain* and regard as cold, because we usually notice the Moon in the chill of night. Thus we think of blue as a 'cold' colour. Makers of lamps and fluorescent tubes, like interior designers, know how important is the psychological response to colour. Even on a national basis, colour matters. Is it a coincidence that the people of so many northern European and northern American countries, with long moonlit nights, have chosen cold blues and whites for the colours of their national flags, while people of the sunnier African, Mediterranean and Central American countries tend to have chosen warm yellows and oranges?

Fig. 25. *Minnaert's observations of 'colour scale' correlate well with photoelectric* B–V *colour index.* ● = *star;* ○ = *Planet.*

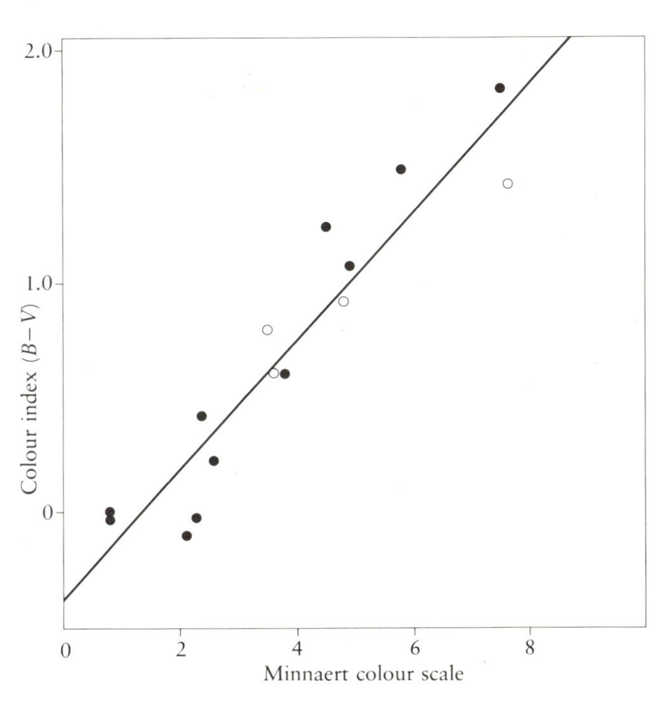

As we have shown in this chapter the perception of colour with the eye is fraught with physiological and psychological pit-falls, although evidently the eye is excellent at perceiving even subtle colour differences between similar objects viewed in similar ways (like bright stars viewed with the eye). With the camera, colours can be caught, intensified and viewed at leisure to reveal the universe of colour in astronomy. We show how the camera records colour in the next chapter.

Smyth's 'Sidereal chromatics'. To help naked-eye observers to use a common vocabulary in describing the colours of the stars, Smyth published a standard colour chart with which to compare the telescopic views of stars. It was circulated as the frontispiece to the edition of Cycle of celestial objects *published in 1881, the best colour reproduction then achievable with Oxford University Press's late-Victorian printing technology.*

Orion. The Orion Nebula is the most highly coloured (pink) object on photographs of the constellation of Orion. The brightest star, Betelgeuse, shows some orange and contrasts with the stars of the rest of the constellation. Inset, a defocussed picture of the same region of the sky picks out the red star in the Lambda Orionis association (top) and emphasises the numerous blue stars in Orion: the brightest images are saturated and appear almost white.

Chapter Two

Photography and colour

Beginnings

Photography has played an essential part in astronomy. Equally, photography has benefitted from meeting the special needs of astronomers. Why does this well-tried technology continue to thrive in the electronic age, and how does photography reveal the colour of stars and nebulae?

The first astronomical objects to be photographed were, not surprisingly, the brightest in the sky. J.W. Draper (1840) made a daguerrotype of the Moon whose familiar markings could be seen on the image. The Sun's more bland countenance was not captured until 5 years later, when, at the behest of Arago (1855), Foucault and Fizeau produced a daguerrotype which showed a few sunspots. Some of the brightest stars were also photographed in this period.

If this achievement seems unremarkable today we must remember that the daguerrotype process was slow, inconvenient, unpleasant and often downright hazardous to use and that the plates were very insensitive to light. As an example, a portrait photograph taken in good conditions would require an exposure of at least 1 min, sometimes much longer, compared with 0.01 s today. This no doubt explains the rather rigid poses of the mid-nineteenth-century worthies who subjected themselves to the 'taking of a likeness'. It certainly explains why astronomers were disappointed with daguerrotypes and considered the results to be not worth the trouble, which was considerable. A daguerrotype plate was made by exposing a newly burnished silver-coated plate of copper to bromine or iodine vapours. These corrosive and very unpleasant elements reacted with the silver at room temperature to produce a thin, light-sensitive layer of silver halide on the polished surface. After a lengthy camera exposure the unseen latent image was made visible by immersing the plate in the toxic fumes given off by a heated tray of mercury. The mercury adhered to the exposed silver halide to form a shiny amalgam which had a positive silvery reflective image. The photograph was made permanent by dissolving away the unused silver salts with a solution of hypo (sodium thiosulphate). This remarkable process enjoyed great commercial success and combined many of the ingredients of modern photography – a light-sensitive layer of silver halide, some form of development to amplify the effect of light on this layer, and a fixation stage to make the image permanent.

It was, however, not well-suited to scientific photography because the images were unique and could not be easily duplicated, they could not be enlarged, they were laterally reversed, and their reflective surfaces made them difficult to view. Above all, the plates were slow and limited astronomical photography to the brightest objects in the sky.

The next few decades were to see many improvements in the science and technology of photography. The daguerrotype gave way to the 'wet collodion' process which, while an improvement in some respects (it gave a transparent negative, for example), still required that the photographer sensitise his own plates as he needed them and expose them before they dried out. In the late 1870s there appeared a new process, the 'dry gelatin' emulsion, a system essentially like the one in use today and for the first time the photographer was freed from carrying his own preparation and processing laboratory with him on his travels. He could purchase his sensitised materials ready-coated and expose and process them at leisure. Emulsions coated on rolls of flexible film were marketed by George Eastman in 1889, thus beginning the modern age of popular photography. For the first time emulsion making was undertaken in research-based commercially oriented companies and the concept of emulsion speed rapidly assumed an importance in the market place.

Among the first to exploit the convenience and sensitivity of the new techniques was the English amateur astronomer Ainslee Common (1883) who made a telescopic picture on the newly introduced dry gelatin plates which for the first time showed objects at the limit of what could be seen. Common had been thrilled by photographs of the Great Comet of 1882 made by the Astronomer at the Cape of Good Hope, David Gill. Gill (1882) mounted an f/4 portrait camera on the same mount as a 6-inch telescope; one of his many practical contributions to astronomy was to improve the drive mechanisms of telescopes so that they smoothly tracked the stars in their diurnal motion. He obtained photographs not only of the Great Comet itself, but of numerous stars too. He remarked on the difference between the magnitudes of the stars seen on his blue-sensitive photographs and their visual (i.e. green) magnitude, discovering, though he did not know it, the means by which colour indices of stars were to be measured, and by which most of the colour photographs in this book were to be made. The comet pictures caused a sensation at the December 1882 meeting of the Royal Astronomical Society where Common saw them for the first time. Inspired by them, Common (1883) photographed (Fig. 26) the Great Nebula in Orion, with his home-made 3-foot telescope located in the garden of his home at Ealing in London. This telescope later became the basis for the Lick Observatory Crossley reflector. On the photographs, with exposures between 37 min and 1 h, Common could see stars which could only be glimpsed in the very best conditions, and the existence of which the keenest-eyed astronomers had disputed.

Common's picture of the Great Nebula won for him the Gold Medal of the Royal Astronomical Society (Stone, 1884). In presenting the medal, the Society's President made it clear that

> *the Society (had) been less influenced by the originality of the methods adopted than by the great practical success which has attended Mr Common's efforts in a most important and interesting field of astronomical research.*

The RAS obviously knew a good man when they saw one, for immediately after the presentation of his medal the Society elected him Treasurer! However, the important feature of Common's work, recognised both by the President in his address and by Common himself, was that the growth of astronomy was no longer limited by the sensitivity of the eye. The perceptible Universe had suddenly increased in size, and photography was established as an indispensable astronomical tool, a role which continues to the present day.

The uncertainties of the weather and the difficulty of finding a suitable telescope preoccupied Common. So did the low surface brightness of the nebulae. He was doubtless affected by another problem, reciprocity failure in the emulsion which he used, but was saved from the last practical difficulty of fogging by the night sky, a problem which was not to arise until more powerful telescopes and more sensitive plates became available. Common thus faced most of the problems still troubling all astrophotographers. About the weather and the telescope we will say nothing, but we shall examine how the other factors affect the way photography is undertaken in astronomy and how modern techniques are used to eliminate or minimise the problems. We shall also discuss new and improved ways of using the photographic images we obtain. First, however, we shall look more closely at the photographic system itself.

The photographic plate

Throughout this book we use the term 'photographic plate' because professional astronomy is one of the few remaining applications of photography where glass rather than film is preferred as an emulsion support.

Originally the reason for this was the instability of film. One of the earliest uses of star photographs was astrometry, where the positions of stars (rather than, say, their magnitudes) were measured with great precision. It was soon found that emulsions coated on a film showed non-linear distortions after processing – the film stretched and shrank haphazardly. While this was too small to be noticed in everyday photography, astronomers (as well as cartographers and precision engravers) demanded the stability of a glass backing for the emulsions which they used. In the last 20 years or so the original cellulose nitrate and acetate bases have been replaced in many applications by a polyester film, which is extremely tough and almost as dimensionally stable as glass, and quite good enough for accurate astrometry. So this reason for the continued use of glass plates in astronomy is no longer valid.

However, astronomers' continued use of glass plates is not just a resistance to change. The focal length of telescopes is large, and to obtain a photograph of astronomically interesting objects the image plane must be large. A nebula 1° across, photographed with a telescope with 12 m focal length, fits on to a film 25 cm in diameter. It is not unknown for astronomical telescopes to have image planes as large as 50 cm square. It is no trivial matter to support

a film of this size by the edges without it sagging in the middle, although it can be done by sucking the film onto a supporting back in a vacuum system. To add to the complication, the optical properties of some telescope designs give the best images in a curved focal plane. In the UK Schmidt Telescope flat photographic plates 14 inches square and 1 millimetre (mm) thick are deformed to fit the surface of a sphere 3.07 m in radius during the exposure. This means that the centre of the plate is bent back 5 mm relative to the corners. After the exposure when the glass plate is released, it springs elastically back to its flat shape again. If large sheets of film are distorted like this, they buckle and stretch irreversibly. Often too the special emulsions of

Fig. 26. *Two of Ainslee Common's photographs of the Orion Nebula. The 5 min exposure (left) shows the Trapezium stars. The 37 min exposure (below, same scale) was the first to show stars too faint to be seen by eye and marks the beginning of the photographic era in astronomy. Common made the exposures with a home-made telescope in Ealing using the newly introduced dry gelatin plates.*

astronomy are thicker than normal, and thick emulsions coated on film are awkward because they tend to curl strongly when they dry after processing. Thus astronomers remain faithful to glass plates because they cannot contemplate the fearful alternative of doing anything with a wrinkled, curly film.

Even if they could use film, there are economic reasons for not doing so. The demand for specialised scientific emulsions is small and they are made in relatively short production runs. It is simpler to coat small batches of emulsion on glass than to set up a film-coating plant to make large-format coatings of a dozen or more products.

Emulsion-making is itself an art, as well as a science, and its details are hidden in commercial secrecy. However, the principle of what an emulsion is and how it works is important in our story.

The photographic emulsion

The emulsion, so-called because of its milky appearance, is in reality a suspension of small silver halide crystals in a matrix consisting mainly of gelatin. Both the gelatin and the silver halide crystals are transparent; the turbidity is due to the high refractive index of the light-sensitive particles. The composition of these particles is varied in manufacture, depending upon the intended application, with silver chloro-bromides being used for printing papers and slow films and silver bromides or (more usually) bromo-iodides used for films and faster products. There are in addition often small amounts of other compounds either on or in the silver halide crystals, often present in minute quantities. There might be sulphur-containing compounds, sometimes used with gold salts to improve the emulsion's sensitivity, or adsorbed dyestuffs to extend the otherwise limited spectral range of the silver halides. The thickness of the emulsion coating on film products is a few micrometres; many widely used monochrome and all colour film products are multiple coatings, each of the layers with different properties.

The average size and spread of sizes of the micro-crystals within the emulsion have a profound effect on its characteristics and much of the technology of emulsion-making is concerned with these interrelated properties. The slowest emulsion may contain particles as small as 0.02 μm in diameter. In the fastest, the large crystals may reach a diameter of 2 or 3 μm. The art of emulsion-making consists of understanding and controlling the many variables to produce a photographic material of exactly the desired properties. These properties are not only determined by the intended application but often by the type of equipment which will be used to make the photographs. Most amateurs have to magnify the small images of planets seen through their telescopes in order to photograph them – in effect they use a camera to replace their eye as it looks through an eyepiece; but large telescopes have such long focal lengths that the image from the sky can be directly focused on a photographic emulsion by the objective lens or primary mirror of the telescope itself; just as in using a hand-lens to form an image of a window on to a wall. In effect the whole astronomical telescope functions directly as a camera; the telescope mirror replaces the camera lens. Now until relatively recently astronomical telescopes were very slow, with focal ratios of around f/16 or greater. (The focal ratio of a telescope is its focal length divided by its lens or mirror diameter. Thus a 4-m mirror telescope with focal length 12 m has a focal ratio of 3 and is said to be 'f/3'. The smaller the focal ratio, the brighter the image formed by the telescope, so large focal ratios, like f/16, are said to be slow relative to small focal ratios like f/3.) The dim images formed by slow focal ratio telescopes made long exposures inevitable. Thus from the beginnings of astronomical photography, emulsions with high speed were considered essential and the major photographic manufacturers, particularly the Eastman Kodak Company, designed and produced a range of plates specifically for long exposures at low light levels. The microscopic structure of such an emulsion, Eastman Kodak type 103*a*O, is shown in Fig. 27(*a*). The electron microscope reveals a very wide range of particle sizes, with a maximum around 3 μm.

A silver halide grain is developable only when it has absorbed a certain minimum number of photons. What this number is need not concern us here, and for the sake of this discussion we will assume that each of the grains in Fig. 27(*a*), irrespective of its size, requires the same number of

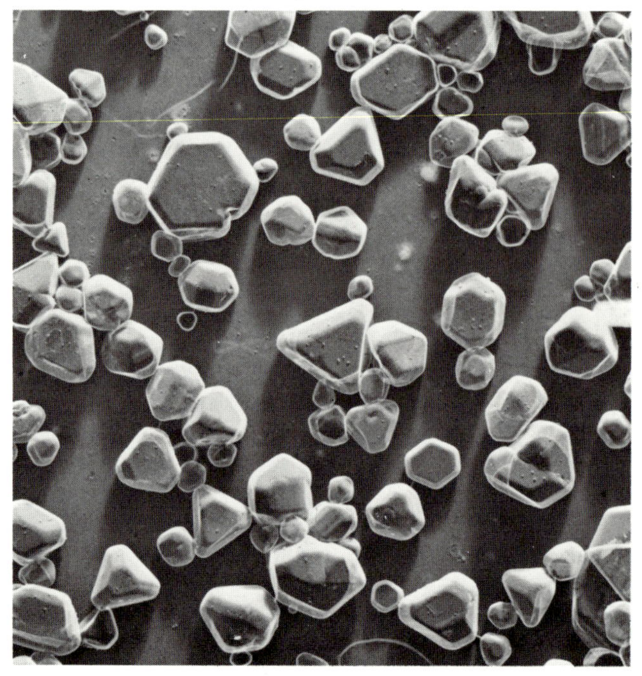

photons. Obviously the larger grains have a greater 'catchment area' than do the smaller ones and will receive their full quota of photons first. These very large grains make the emulsion very sensitive. The large crystals of silver halide are transformed by the developer into even larger filaments of metallic silver and thus the developed image is very grainy, a normal characteristic of fast emulsions. The smaller crystals in Fig. 27(*a*) require a longer exposure to produce rather less silver, so an emulsion of this type continues to increase in developed density over a very wide range of exposure levels (or times). Type 103*a*O is therefore considered a relatively low-contrast emulsion with high speed, coarse grain and low resolution.

In the early 1960s astronomers began to appreciate that a rapid rate of blackening (i.e. high speed) was not the same thing as the rapid acquisition of information and that ultimately it was information, not speed, that interested astronomers (Marchant and Millikan, 1965; Hoag, Furenlid and Schoening, 1978). This realisation coincided with the design and construction of a new generation of large telescopes

Fig. 27. *Astronomical emulsions. Transmission electron micrographs of shadowed replicas show the unexposed silver halide crystals of photographic emulsions designed specially for astronomy.* (a) *(facing page) shows a wide particle size range with some very large crystals, typical of a fast, grainy, relatively low-contrast emulsions, Eastman Kodak type* 103aO. (b) *(above) is a similar image of the newer, fine-grain high-contrast emulsion type IIIaJ now preferred for photographing the faintest objects.*

which featured quite short focal lengths and very fast optics (around f/3). The images produced by these telescopes, although very bright (f/3 is photographically a factor of 28 faster than f/16) were very small, and it became clear that plates of high resolution would be needed to exploit these new instruments to the full. A thorough reappraisal of the photographic needs of astronomers led to the introduction of a new type of photographic emulsion specifically for recording faint objects, an emulsion first used on a telescope by Allan Sandage and Bill Miller (1966).

The special emulsion was the result of close collaboration between the astronomical community and the Eastman Kodak Company, which has for many years catered for the needs of this relatively small but highly specialised market. The search for the ultimate photographic detector is of advantage to both parties; the emulsion-makers explore the limits of their craft while the astronomers discover a larger Universe.

The emulsion was later designed IIIaJ. IIIa is the Eastman Kodak code for the basic emulsion type. Suffixes J, F, etc., indicate the spectral sensitivity conferred on the basic emulsion by the addition of minute amounts of sensitising dyestuffs. The microstructure of IIIaJ is shown in Fig. 27(*b*) to be quite unlike its earlier counterpart. If we once again assume that all the crystals have a similar sensitivity threshold then these uniformly sized grains will become developable at much the same exposure level. Such an emulsion has very little exposure latitude and is of high contrast and resolution, but these grains are also rather small, which suggests that a lot of light is necessary to expose them, i.e. that the speed of the emulsion is quite low, an obvious disadvantage in astronomy, even with a fast modern telescope. While that is certainly true, we should appreciate that photographic sensitivity does not depend on the construction of the emulsion alone.

Reciprocity failure

The photographic effect of light on an emulsion depends not only on the total number of photons received by each silver halide grain, but also on their rate of arrival. This effect is not normally apparent in everyday photography where exposure times are rarely longer than 0.1 s or shorter than 0.001 s. Light levels which make it necessary to use exposure times outside this range, for example the low light levels of astronomy which entail using long exposures, lead to a breakdown of the reciprocal relation between exposure time and illumination. The reciprocity relation expresses the fact that the light energy in an exposure of 0.001 s to 1000 units of brightness is the same as an exposure of 1 s to 1 unit of brightness. However, the photographic effect of the longer exposure at the lower light level is much less. We can express this by saying that the speed of the emulsion drops for long exposures. As an example, a normal black and white camera film (Plus X) exposed for 0.01 s has a nominal design speed of 125 ASA. According to Eastman Kodak (1973) the effective speed of Plus X falls to only 25 ASA when the exposure is increased to 10 s and to a mere 10 ASA when an exposure time of 100 s is needed, a result typical of a film not specifically designed for long exposures. This departure from a fixed relation between exposure time and photographic effect is known as reciprocity failure. More accurately it is low-intensity reciprocity failure, as there is a separate effect for high-intensity sources. In what follows we refer always to low-intensity reciprocity failure. It has been the bane of astronomical photographers since the time of Draper and Common, and has a particularly severe effect on colour photography. Under the leadership of C.E. Kenneth-Mees, the Eastman Kodak Company developed a special set of emulsions for long exposures and gave them numbers which included the letter 'a'. Astronomers widely believe that 'a' stands for astronomy, but we haven't been able to confirm this. Reciprocity failure in these emulsions is minimised.

The mechanism of reciprocity failure has been the subject of extensive research, in part because its nature reveals a good deal about the fundamental processes of photography (Hamilton, 1977). Each crystal of silver halide in the emulsion acts as a solid-state detector of radiation, operating quite independently from its neighbours. When a quantum of radiant energy (a photon) is absorbed by the silver halide grain, a photoelectron and a positively charged 'hole' are produced. The electron and 'hole' are mobile within the crystal lattice until one or both are trapped or they recombine. The trapped electron is capable of neutralising a silver ion, forming a single silver atom which by itself is rather unstable. The single silver atom may soon be lost by recombination with a 'hole' or by reaction with impurities on or around the crystal. If, however, another photon quickly appears to liberate another silver atom the two atoms may combine to form a stable but undevelopable pair, a sub-latent image centre which can grow, with the addition of more silver atoms, to a latent image capable of triggering the development of the crystal.

This mechanism explains why the photon arrival rate affects photographic speed and suggest several ways in which the inherent sensitivity of the system might be improved, for instance by providing more centres where photoelectrons might be trapped.

Hypersensitisation

Techniques which minimise or eliminate reciprocity failure have been comprehensively reviewed by long-time astrophotographers Alex Smith and Art Hoag (1979) and by Tom Babcock (1976). These methods, which may at the same time also give an increase in photographic speed, are collectively known as *hypersensitising*, or *hypering* for short. Almost all observatories which practice astronomical photography use hypersensitisation in one form or another and in fact hypering is essential for some modern photographic emulsions if they are to be used at all. Not only do these newer materials suffer from reciprocity failure but they are inherently slow to start with.

It has been found that one of the main causes of reciprocity failure is oxygen and water naturally present in the gelatin of an unexposed emulsion layer (James, 1972). These are contaminants. They can be quite simply sucked out by a vacuum or flushed away in an inert gas such as nitrogen. An even better result is obtained if the plates are baked in an oven (65 °C) for a few hours in nitrogen gas. Apart from removing the contaminants, baking also extends the chemical sensitisation begun by the manufacturer. A further large gain in speed and complete removal of reciprocity failure is obtained if the baked plates are soaked in hydrogen gas for several hours at room temperature (Babcock *et al.*, 1974). This apparently hazardous and somewhat bizarre practice is a gentle kind of chemical reduction process which 'seeds' many of the silver halide crystals with silver atoms to form ready-made sites where photon-induced atoms can form stable latent image specks. One of the limiting factors in further hypersensitising is the growth of chemical fog. Grains which accumulate too many silver atoms spontaneously become developable even without the trigger of a photon or two.

These now quite elaborate procedures have evolved over the last 10 years or so, though the first use of baking seems to be due to Bowen and Clark (1940). To obtain optimum results, all types of emulsion, and often individual batches of a given emulsion, require their own special recipe. When hypered, plates have a short shelf life, perhaps just a few hours (which is why plates are not hypered by the manufacturer). The plates must be prepared just before use and are stored and exposed in a dry nitrogen atmosphere and developed as soon as possible after exposure.

It is extremely fortunate that the newer, fine-grained but rather slow emulsions, designed specifically for astronomy, respond so well to hypersensitising procedures which, without altering their desirable properties, make them fast enough to be astronomically useful. With these products, the 1.2-m UK Schmidt Telescope at Siding Spring in Australia has just completed a deep photographic survey of the southern sky, with each of more than 600 survey fields (and many non-survey projects) being recorded on hypered IIIa plates in 60 min or less. A similar survey of the northern sky is planned. Neither survey would be feasible without hypersensitisation.

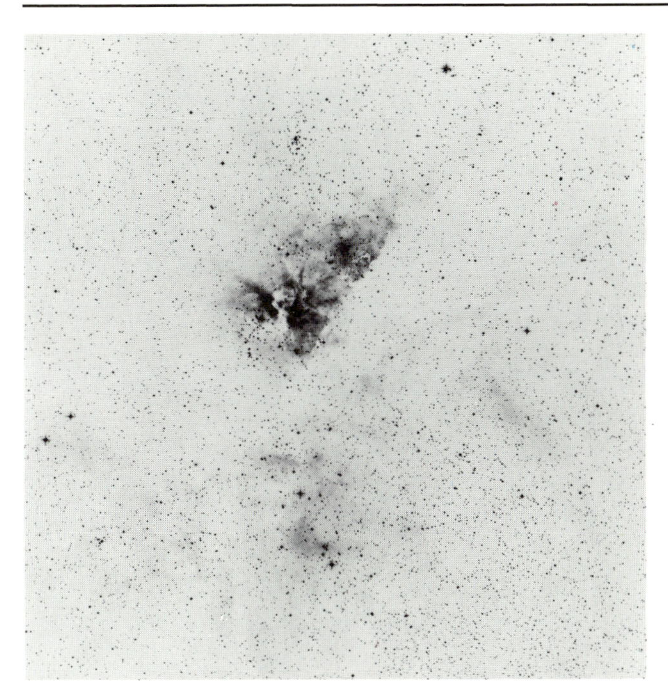

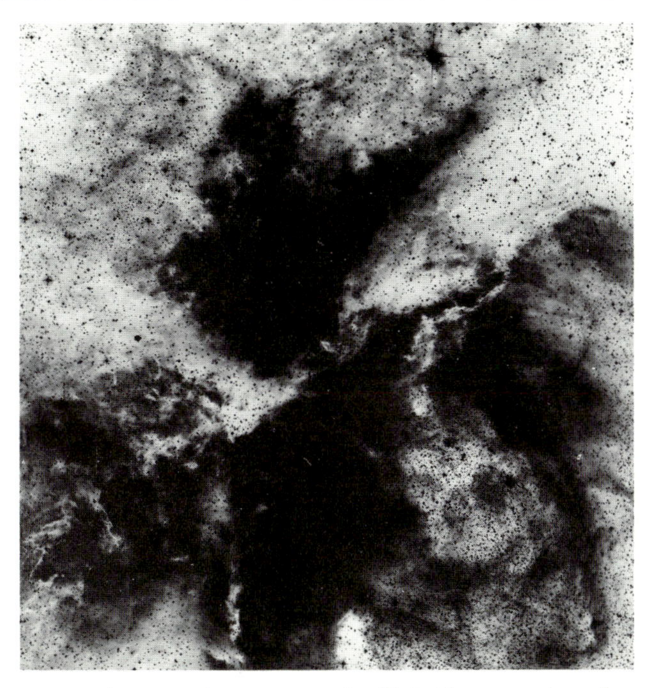

Fig. 28. *Effect of hypering. These photographs of the Eta Carinae Nebula were taken under identical conditions. They used the same emulsion and filter, the same telescope and exposure time (10 min) and they were made on the same batch of IIIaJ emulsion within a few minutes of each other. The only difference is that* (b) *(right) used a plate which had been hypersensitised in nitrogen and hydrogen, whereas* (a) *(left) was exposed as received from the manufacturer. Hypering is indispensable in photographic astronomy.*

An indication of the improvement in effective speed due to hypering is shown in Fig. 28 which shows two exposures of the Eta Carinae Nebula taken on IIIaJ plates with the UK Schmidt Telescope. The pictures were taken under identical conditions: filter, exposure time, emulsion type, even emulsion batch. The picture on the right, however, was taken on a plate which had been hypered in nitrogen and hydrogen, while the one on the left was exposed on a plate as received from Eastman Kodak. With careful technique, speed gains of 10 or 20 times can be achieved which enable these excellent materials to be used in circumstances where previously only their coarse-grained predecessors were considered suitable.

Light from the night sky

Apart from the distant stars and galaxies in which astronomers are interested, the highly sensitised plates are also capable of registering other, unwanted sources of light. But where does this light come from? In a dark place away from the cities, such as an observatory site on a mountain there are no street lights. However, energetic particles, travelling out from the Sun, enter Earth's atmosphere and cause it to glow faintly. Extreme examples of this, seen when the Sun is very active, are the aurorae, the flickering curtains of colour seen in both northern and southern hemispheres from high latitudes. Even when the Sun is quiet, electrically charged particles bombard our atmosphere day and night producing a faint light called the *airglow*. Other sources of light are sunlight scattered by the dust in our solar system (the *zodiacal light*) and starlight scattered by dust between the stars of the Milky Way as well as stars too faint or too numerous to see individually. So even on the darkest night the sky is never truly black. On a fast telescope therefore photographic exposures are limited not by the time available but by the fogging of the plates by the faint glow of the night sky.

Now we can see why high contrast was an important design consideration in the IIIaJ emulsion. The object of interest might be a star or galaxy which adds only slightly to the night sky brightness. When contrast is high a very small increase in exposure level will give a measurable increase in blackening on the processed plate, an increase which might go undetected if the contrast was lower. In other words the plates are exposed so that their output signal-to-noise ratio is at a maximum, noise in this context being the uniform density due to the night sky, together with that from non-image fog, mainly due to hypersensitising processes. On the 3.9-m Anglo-Australian Telescope (AAT) and the UK Schmidt Telescope, both with focal ratios around f/3, such 'sky-limited' plates are obtained with exposures of 60–80 min on hypered IIIaJ plates. The exposure time is critical: too short and faint objects are not recorded; too long and the background becomes too dense and grainy to give the best results.

Fig. 29. *Newton's experiments on spectra as depicted in Voltaire's* Elements de la philosophic de Newton, *Amsterdam, 1738. Sunlight passing through a hole in the window is dispersed by a prism into a spectrum thrown onto the paper, P.*

Trichromaticity

Developed silver, deposited in a photographic emulsion by the image of a bright object, appears black. A photographed image is thus reversed to a negative. If the photographer rephotographs the negative, as he does when making a print, the image is reversed once more, to a positive, so that bright objects appear white. No matter how accomplished the photographer, the picture is still 'black and white'. It is remarkable that our brains make any sense of it at all.

Dissatisfaction with the appearance of black and white images led early to a search for even more natural photographs which showed colour. But astronomers were slow in using colour photography for their own science, so the history of colour photography in astronomy is short. The first colour pictures of objects outside the Solar System did not appear until 1959, almost 100 years after the invention of colour photography itself; but the foundations on which astronomical colour photography is built were laid at the time of the Great Plague of 1666. Among the refugees who fled the urban disease was Isaac Newton, displaced from his rooms in the University of Cambridge to his family home on a farm near Grantham in Lincolnshire. In the relative safety of the English countryside Newton was led to consider the nature of light (Fig. 29). He was able to show with ingenious experiments with simple pieces of glass that sunlight could be split up into its component colours from 'the outmost violet-making rays' to the 'utmost red-making rays', with blue, green, yellow and orange in between. He was also able to show that this spectrum could be recombined to produce white light and that the removal of one or more parts of the dispersed spectrum gave a light which was no longer white. Thus in a few short weeks in the summer of 1666 he laid the foundations of spectroscopy, colorimetry, optics and colour science in general. For all his insight on the physical nature of light and colour, Newton had little to say about the physiology of the way we actually perceive the colours of the natural world. Newton's results on *Optiks* were published in 1704. As described by Walls (1956) and MacAdam (1970) it was not until 1777 that George Palmer, a forgotten pioneer of optical science, produced a recognisably modern theory of colour vision, with its recognition of the importance of three colours:

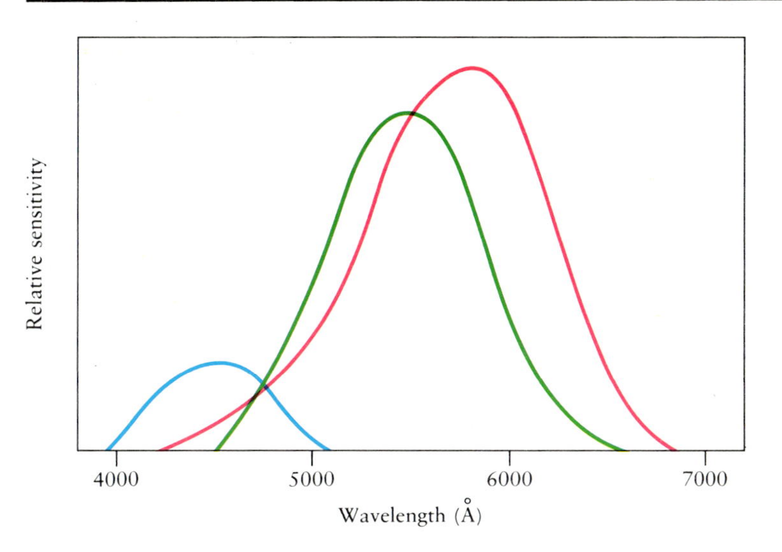

Fig. 30. *Spectral sensitivity of the eye. The three colour receptors of the eye have the relative sensitivities to light which are shown in this figure. Notice the relatively low blue sensitivity, and the difficulty of red–green discrimination because the sensitivity curves overlap. Nonetheless the eye evidently sees colour! Contrast these spectral sensitivity curves with those of colour film in Fig. 34.*

> *Each ray of light is composed of three rays only. One appears yellow, one red and the other blue . . . The surface of the retina is composed of particles of three different kinds, corresponding to three rays of light . . .*

This is remarkably similar to the ideas of Thomas Young (1802). It is Young who is normally credited with the concept of trichromatic colour vision which, after elaboration by von Helmholtz (1860), became known as the Young–Helmholtz theory of colour perception.

According to this theory there are three kinds of colour receptors in the eye; they are located in the cone-shaped cells of the retina. Three types of cones are sensitive to three overlapping bands of colour – blue, green and red (Fig. 30). The three receptors may be stimulated in any ratio by a spectrum of light and this produces the sensation of colour in our minds. For instance, stimulation of the red and green receptors by equally bright red and green lights produces the sensation of yellow. A single spectral emission line near 5800 Å is also perceived as yellow because it stimulates the red and green receptors in the eye equally, and the blue receptors virtually not at all.

Note that the blue response of the eye is weaker than the green and red responses. The relative difficulty which Struve and Smyth had in differentiating blue and white stars has already been remarked upon (p. 20). This difficulty is partly because the spectra of hot stars are relatively flat and uniform, but even so stars with a temperature of 30 000K should look as blue as the sky (Fig. 107). They do in fact photograph blue (Fig. 37). They don't look as blue as the sky because, when the eye looks at faint stars, the blue-sensing receptors in the retina are activated with more difficulty than the green and red ones. All the colours of the spectrum from violet to red can be mimicked by combinations of different intensities of three standard wavebands in the blue, green and red. The range of spectral colours does not exhaust all the possible combinations of these three stimuli. For instance, equal blue and red is seen as purple; there are no spectral lines which seem to be this colour. A range of colours produced by mixing three lights is shown in Fig. 31.

We must not think that this theory of colour vision is the whole story. Colour mixtures formed by three lights are processed by the brain; in real life, the brain perceives also the texture and form of coloured objects. The Young–Helmholtz theory cannot reproduce grey or silver or gold, for instance, and these colours do not appear in Fig. 31; nor can it produce brown. Moreover, Edwin Land has recently demonstrated how colour perception can be stimulated by only two colours rather than three if the colour mixtures are of objects rather than patches of light. Clearly colour vision is a complex psychological subject. Nonetheless, using the Young–Helmholtz theory of trichromatic colour vision as a spur to his work, James Clerk Maxwell first demonstrated the practicality of colour photography.

Fig. 31. *Range of colours from the three stimuli. This is a photograph of a painting made to illustrate how three different coloured lights can in principle be combined to reproduce a range of colours. Any combination of the three lights at three different intensities is represented by a point in the overall triangular shape. Equal intensities yield white (centre). The colours of the spectrum can be represented around two edges of the triangular shape. The third edge shows non-spectral colours such as mauve. The palate of colours available in any three-colour process is a restricted subset of this theoretical diagram.*

Maxwell's fame rests largely on his work on electromagnetism and the kinetic theory of gases, but, like Newton, he was intrigued by the laws of colour mixture. His studies of colour led to what we would now call a conceptual model of the mechanisms of colour vision. It was to explore this model that Maxwell made the experiment which was the first demonstration of the feasibility of three-colour photography.

Colour photography

Maxwell's experiment had a profound effect on all subsequent work on colorimetry, sensitometry and colour vision, although alternative methods of colour photography eventually replaced his rather inconvenient additive process. This system has, however, been found to be ideal for the unique requirements of colour photography in astronomy and has been revived and improved with modern materials to produce many of the pictures in this book. We shall look more closely at the train of thought which led to Maxwell's 'invention' of colour photography.

Maxwell was obviously familiar with Young's theory (although, like everyone else, he seems to have ignored Palmer's work, 25 years before Young). Maxwell (1861) wrote:

> *Each nerve acts, not by conveying to the mind the length of an undulation of light, or of its periodic time, but simply by being* more *or* less *affected by the rays which fall on it.*

In the same paper he described how 'the art of photography' might be used to recreate a coloured scene. Anticipating the invention of the panchromatic emulsion by over 40 years he said:

> *Let it be required to ascertain the colours of a landscape by means of impressions taken on a preparation equally sensitive to rays of every colour. Let a plate of red glass be placed before the camera, and an impression taken. The positive of this will be transparent wherever the red light has been abundant in the landscape, and opaque where it has been wanting. Let it now be put in a magic lantern along with the red glass, and a red picture will be thrown on the screen. Let this operation be repeated with a green and a violet glass, and by means of three magic lanterns let the three images be superimposed on the screen. The colour of any point on the screen will then depend on that of the corresponding point of the landscape, and by properly adjusting the intensities of the light, etc., a complete copy of the landscape, as far as visible colour is concerned, will be thrown on the screen. The only apparent difference will be that the copy will be more subdued, or less pure in tint than the original. Here, however, we have the process performed twice – first on the screen, and then on the retina.*

Maxwell used the technical expertise of an experienced photographer named Thomas Sutton to make a series of negatives of a collection of coloured ribbons. Each plate was taken through a blue, green, yellow or red filter, separating the scene, as it were, into its constituent colours. The colour separation negatives were made into positive transparencies. Before a distinguished audience at The Royal Society in London, Maxwell (1861) demonstrated the world's first colour photographs. Combinations of the positive transparencies were projected in register through their appropriate filters. According to Sutton (1861), when these different-coloured images were superimposed upon a screen, 'a sort of photograph of the striped ribbon was produced in the natural colours'.

It is actually quite remarkable that any kind of colour picture appeared at all since the separation negatives were made by Sutton on blue-sensitive plates which were quite insensitive to yellow, green or red light! Serendipity played an essential part in making the demonstration a success – just how fortunate Maxwell was has been shown by Evans (1961*a*,*b*) who found that the red ribbon was also highly reflective in the ultraviolet which was transmitted by the somewhat imperfect red filter. Thus the red ribbon was actually recorded by means of ultraviolet light! It was Vogel (1873*a*,*b*) who discovered that dyes could be used to extend the sensitivity of photographic emulsions, first to green and later to the red regions of the spectrum, thus making Maxwell's experiment generally possible and removing the luck from it.

The ephemeral nature of the projected image was an obvious limitation to widespread use of the process but it was not long before many devices for both taking and viewing positive colour separations were available. Many of the possible ways of making colour pictures were described by Ducos de Hauron (1869) and Charles Cros (1869). These two Frenchmen were the first to discover a practical system for making colour prints by means of a subtractive process. They proposed that colour separation negatives should be made along Maxwell's principles through red, green and blue filters. The red separation negative would be used to make a blue-green (cyan) positive, the green negative a blue-red (magenta) positive and the blue negative a yellow positive. The positives would either be made separately on plates, and subsequently superimposed to give a transparency, or the positives would be printed in register either as dyes or pigments (inks) on to a white surface to yield a print. These ideas formed the basis for almost all of the subtractive colour photographic processes in use today.

At the end of the nineteenth century the additive process was still in regular (though not widespread) use for making colour pictures. It was typified by lenticular-screen or dyed-grain processes introduced by Joly (1894) and by the Lumière brothers' Autochrome system of 1907, but these were not satisfactory. They produced rather dark transparencies and it was difficult to make good copies or paper prints. The plates were of very slow speed. Nonetheless processing was fairly straightforward and these additive methods survived until the outbreak of the Second World War. We know of no record of their being used in conjunction with a telescope, astronomical or otherwise.

Subtractive colour picture making became dominant. A variety of methods were in use, all of them involving at some stage three distinct colour separation negatives. In 1935, years of experiment and development culminated in the integral tripack film where the three separation negatives are combined in a single multilayer emulsion (Davies, 1936). No longer individually accessible, the colour separations remain in perfect register without the photographer being aware of their existence.

It is surprising, in view of the skill and imagination of the early colour photographers, that no-one attempted to make colour separations of astronomical objects. All the ingredients were available – colour-pure images from achromatic reflecting telescopes, colour filters, multispectral plates and methods of combining them – yet no-one put them together to excite the imagination of astronomer and layman alike. From Maxwell's Royal Society demonstration it was almost 100 years before the first colour pictures of astronomical objects appeared. These early photographs were made on a conventional tripack colour reversal film.

Direct use of colour films

The first serious attempts to use conventional colour films for long-exposure photography of objects outside the Solar System were made by the late William C. Miller (1959), then Research Photographer at the Hale Observatories. As a result of many preliminary experiments and considerable patience at the 5-m Hale and the 1.2-m Schmidt telescopes on Mt Palomar he was able to produce the first of what were to become an extremely comprehensive set of pictures, which are still available from the California Institute of Technology Book Store. The introduction of 'fast' colour films in the late 1950s made telescope exposures short enough to be practically useful. However, reciprocity failure took a heavy toll of this new-found speed and Miller's (1959) picture of the nebulosities in the Crab Nebula required an exposure of 4 h at the f/3.6 prime focus of what was then the world's largest telescope (Figs 32, 33).

Fig. 32. *(above)* *Crab Nebula as photographed. In 1958 W. Miller used the 200-inch Palomar Telescope in an exposure of 4h on colour reversal film (Super Anscochrome) to obtain this photograph, reproduced here directly from the original. The colour balance is seriously distorted by reciprocity effects, which make the photograph far too blue. The red, particularly in the filaments which cross the diffuse central nebula, is suppressed because the red-sensitive layer in the colour film has lost relatively more sensitivity than has the blue layer. Nonetheless the information is still there and, as in the next figure, the filaments can be enhanced by rebalancing.*

Fig. 33. *(right)* *Crab Nebula (colour balanced). After taking the photograph in Fig. 32, Miller rebalanced it by printing through colour filters designed to emphasise the spectral bands which most suffered reciprocity failure. This process restored the red colours to the wispy filaments. It ought to have made the central diffuse light in the nebula white, but the central light in fact has a yellow cast because the relative contrast of the three layers in the original film was upset by the long exposure. This reproduction was made directly from Miller's rebalanced original.*

The material which Miller seems to have preferred for his original work (Super Anscochrome) had a camera speed of 100 ASA for an exposure of 0.01 s, but fell to 7 ASA during a 4 h exposure! (Miller, 1962). The poor long-exposure speed of these early colour films was by no means the only difficulty they presented. The characteristics of the three layers of a colour film are carefully chosen to give accurate colour balance over the limited range of exposure times found in normal practice, typically 0.001 – 0.1 s. Exposures significantly longer or shorter than this give rise to changes in sensitivity which vary from layer to layer – each layer suffers reciprocity failure in its own way.

To quote Miller (1959):

> *In such faint light, photographic film works less efficiently than in bright light, and in color film the red- and green-sensitive layers of the emulsion lose much more sensitivity than the blue layer. To bring the Crab's colors back into balance, a copy had to be made from the original, filtering out two-thirds of the blue and a fifth of the green.*
>
> *Such intricate color-correction processes, worked out with the co-operation of Ansco engineers during two years of experiments, assure that the pictures you see here are as true as the photographic art now permits.*

The complicated filtration corrections which Miller used could only compensate for changes in sensitivity of the three layers. Reciprocity also affects the contrast of a photographic film and again each layer reacts differently. However, contrast shifts are a much more serious problem. A contrast shift may make bright objects look too blue, but faint ones too red. Viewing the transparency through a blue filter may correct the faint objects but makes the bright ones look worse. There is no simple cure for this 'crossed curves' condition in a colour film but its effects can be minimised by making the original exposure through substantial colour correction filters. This increases the exposure time, which is already too long.

The filtration correction techniques and careful colour sensitometry used by Miller, together with much practical advice for those contemplating long-exposure colour photography, are given in a photocopy publication *Color photography in astronomy; a discussion of problems and solutions*, initially available from Miller at Hale but out of print since his retirement. The originals for the standard set of Palomar colour pictures, widely duplicated (with varying degrees of success) for sale throughout the world, were used to make the early colour pictures in publications by Miller (1959, 1961, 1962). Later copies, including those available today, seem to have lost much colour saturation and some contrast, possibly by being made from several generations of copies now remote from the master transparencies. Nonetheless, Miller's original results are difficult to better with modern materials, although exposure times are now somewhat shorter.

Miller's work stimulated a good deal of investigation into ways of making low-light-level colour pictures which overcame or bypassed the problems of reciprocity failure. Some of these methods will be described later. In spite of their low efficiency, conventional tripack colour films have continued to be used in astrophotography, mainly because they are convenient to handle. Most workers seemed to prefer reversal (i.e. transparency) films, but Mitton (1977), Murdin, Allen and Malin (1979) and Malin (1980) have published a number of colour photographs using negatives which have been made using the UK Schmidt and the Anglo-Australian Telescopes at Siding Spring in Australia. Negatives have the advantage that moderate over- or under-exposure can result in a usable photograph which would be lost on slide film and correction of most colour balance problems is fairly straightforward.

Cooled cameras

Since Miller, many people have shown that colour astrophotography is a practical proposition, given fast optics and the facilities for experimentation. Another important ingredient is patience, for, as we saw earlier, photographic speed falls dramatically with very long exposures. Furthermore, long exposures affect the overall colour balance of films because each layer has a different degree of reciprocity failure. Instead of compensating for shifts of colour balance caused by long exposures, however, it might be possible to do something about the causes of reciprocity failure itself.

Temperature is one of the factors which affect the growth of the latent image in a photographic emulsion. When the photons arrive slowly, thermal effects in the silver halide crystals may destroy the latent image as fast as it is created. Hence cooling the emulsion during exposure might help. Webb (1935) recognised this quite early in the growth of photography, and this approach has been used occasionally in astronomy. Hoag (1960) seems to have been the first to apply the technique to colour films, as an extension of his prescient investigation into the use of fine-grained monochrome materials in astronomy. He found that the gain in effective speed and stability of colour balance obtained was useful. It justified the construction of a special cold camera with a heated window and vacuum chamber to prevent frosting. With a second camera and more complete investigation Hoag (1961) found a 'most spectacular' increase in response of High Speed Ektachrome exposed at $-78\,^{\circ}$C for 60 min. Similar results for High Speed Ektachrome were obtained by several other workers. Hoag also found that the colour balance of the film was unchanged during the long exposure.

The practical difficulties involved in the construction and operation of cold cameras have generally limited their use to the smaller formats, usually 35-mm films. No-one seems to have made a cold camera to accept the larger films needed to exploit this technique with a fast, wide-field telescope. For instance, at the prime focus of the Anglo-Australian Telescope, plates 10 inches square are used to cover a field 1° in diameter. This is too large an area to cool uniformly without considerable difficulty. In the UK Schmidt Telescope, plates 14 inches square are used inside a camera system where access to the focus is severely limited. It is completely impractical to cool such a large area in so confined a space. The use of commercial colour films for long exposures is thus beset with practical problems which are difficult though not impossible to resolve. Another limitation, inherent in all tripack colour materials, is not so easily remedied by any of the approaches described above: it is *colour balance*.

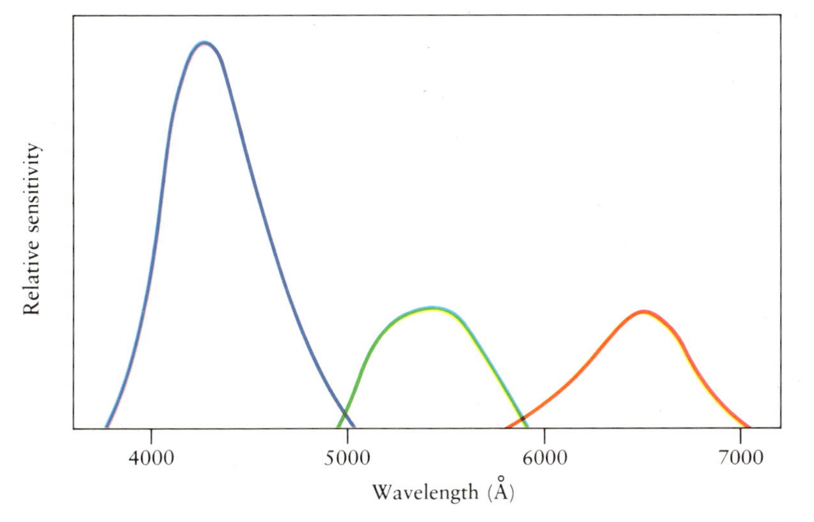

Fig. 34. *Spectral sensitivity of colour film. The blue-, green- and red-sensitive layers are all well separated in spectral sensitivity compared with the eye (compare with Fig. 30). Notice the marked insensitivity at 5000 Å between the blue and green curves, neither of which layer therefore senses the 5007 Å emission line in astronomical nebulae.*

Colour balance

Colour films are designed to picture the world which we see around us. They differ in the degree of realism which they provide – indeed colour film manufacturers have found it profitable to try to satisfy customers rather than reproduce reality. Colour film thus has a brighter-than-life quality so that holidays look sunnier and holidaymakers more tanned than perhaps they did in reality. The physical conditions which colour films render well are relatively uniform areas of the broad band colours of everyday life. (That rather obvious statement belies the complexity of the problem, as is well illustrated in the book by Hunt (1975).) In at least two ways astronomical photography depicts scenes outside these conditions.

First, astronomical scenes are well modulated in intensity – there are very bright areas and very faint areas in the same scene, unlike most everyday scenes which are relatively uniform. If the film is exposed to record the faint parts then the bright parts will be overexposed. Since the amount of any active chemical in a film is limited, the effect of overexposure is to use up completely the chemical – or *saturate* the film. In colour film a sufficiently bright object will saturate all three sensitive layers. It thus appears equally bright in all three colours – i.e. white. It is clear now why star images in astronomical photography are usually white, particularly the brighter stars.

Secondly, astronomical scenes often show light originating in emission lines. Such scenes are unusual in everyday situations, since everyday scenes are usually lit by the continuum spectrum of the Sun. Colour film depicts unrealistically the colours of emission-line spectra because of its spectral response. The spectral response of a normal colour film is the sum of the response of its three constituent layers, as shown in Fig. 34. The three spectral sensitivity curves are the result of a series of practical compromises which must be faced by any manufacturer of colour film. While they match in their gross features the sensitivity curves of the three colour receptors in the eye, they do not correspond in every detail to its colour response. Most photographs are made with reflected radiation from the Sun or filament lamps or electronic flash units, all of which have slowly changing continuum spectra. The marked gaps in the rather uneven spectral response of colour film are

Fig. 35. *These two photographs of the AAT were taken a few minutes apart on the same roll of daylight-balanced colour film. In both cases a casual observer would describe the illumination as 'white light'. In* (a) *(left) the exposure was made by the fluorescent lighting normally used in the dome by day. Picture* (b) *(above) was made with the dome open and thus by sunlight reflected from the afternoon's cumulus cloud. (The fluorescent lights were left on during this daylight exposure, but did not contribute significantly to the illumination of the telescope.) The strong narrow emission at* 5460 Å *in the spectrum of the lamps is near the peak sensitivity of the green layer of the film and the lamps are generally deficient in red light. The green cast is obvious in* (a).

therefore of little consequence. If however, a colour film is required to record radiation which is not broad band but consists of narrow, more or less monochromatic lines, its weaknesses are revealed. A good example is the strong greenish cast of colour pictures taken under fluorescent lighting (Fig. 35(*a*)). To the eye fluorescent lighting is a perfectly acceptable white – even a warm white. The greeness of pictures taken with it is due to some very strong green emission lines from mercury ionised by the electrical discharge within the tube. The ionised mercury emits ultraviolet light which activates the phospor on the tube walls so that they glow white. The unuseful green emission passes through the tube walls and contributes generally to the light from the fluorescent lamp. The green spectral lines hit the peak of the green spectral sensitivity curve of colour film, so this radiation is emphasised by colour film as an excess of green. It can only be corrected with a strong magenta filter over the camera lens, to reduce the intensity of the spectral lines.

Similarly, many of the objects of interest to astronomers (though not usually stars) emit much of their visible radiation in the form of narrow emission lines which miss the peaks of the sensitivity curves of colour films. This is especially true of the gaseous nebulae. Many of these emit strongly in the green lines of doubly ionised oxygen at 4959 and 5007Å. These spectral lines fall in the marked dip in the sensitivity curve of the colour film (Fig. 34) and are not well recorded – indeed this prominent colour is notably absent from all colour film pictures of gaseous nebulae, being overpowered by the red line of hydrogen which falls on a peak of the film's sensitivity curve at 6563 Å (Fig. 36).

To the difficulties of reciprocity failure and non-uniform spectral response in colour film must be added one more – all colour films fast enough to be useful in astronomy have low contrast. To photograph faint objects against the glow of the night sky a contrasty emulsion is essential.

It must be said at once that these limitations, though severe, have not rendered colour films completely useless in astronomical photography. Recent developments in hypersensitising have gone some way to offset the ravages of reciprocity failure, as for example in the experiments by Smith and Schrader (1979) at Florida's Rosemary Hill Observatory. The introduction of high-speed materials which are even faster (and more contrasty) after forced processing is another way in which colour films can be made to give useful results.

As an example, observe in Fig. 37(*a*) the out-of-focus field of Antares (the same area as Fig. 24 in Chapter 1). The stars are identified in Fig. 37(*b*) and their colour indices listed in Table 7. The colours as photographed and measured generally match well. There are peculiarities, and Antares itself is an

Fig. 36. *Spectra recorded on colour film. In the fainter spectra the three dyes of the colour film produce red, green and blue images in the spectrum, which scarcely overlap, giving the impression that each spectrum is made up of three distinct bands of colour instead of gradually changing colour from red to blue. In the brightest spectra yellow shows where the green and red images overlap.*

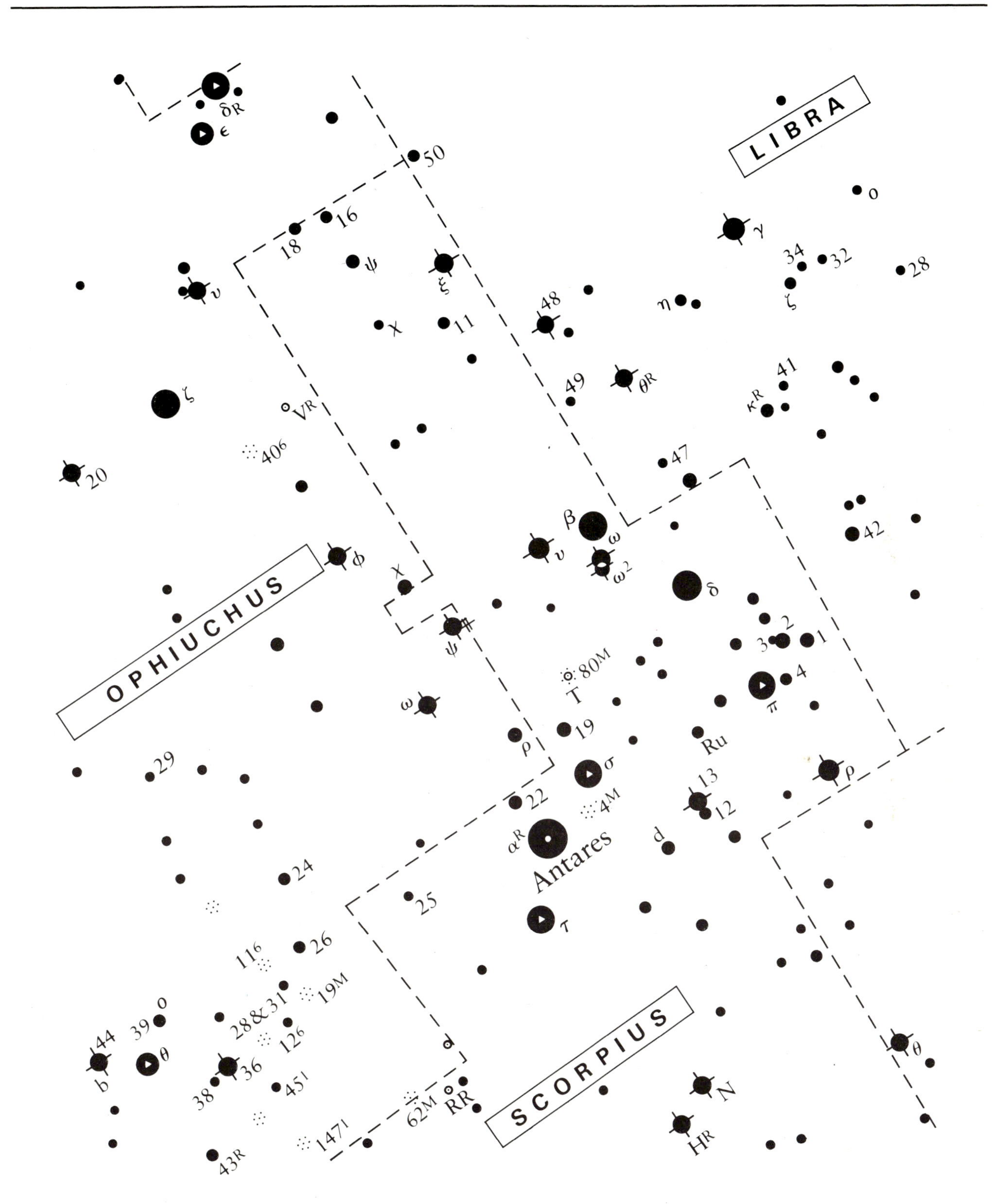

Fig. 37. *Antares.* (a) *(left) This picture, which was taken with a conventional camera with colour reversal (i.e. slide) film, was deliberately made out of focus to spread the star images. Even so the six brightest stars are burned out. The brightest star is Antares at lower centre which shows only faintly its otherwise distinctive red colour. The others show colours indicative of their surface temperatures. The stars are identified in the diagram above* (b) *and the* B–V *colours are listed in Table 7. The same region with a similar exposure but taken in focus appears as Fig. 24.*

anomaly. It should be the reddest star on the photograph but is near-white because each of the three colour-sensitive emulsions is near-saturated and thus almost equally exposed. The white stars generally are the brighter ones, even though their colour indices suggest they should look bluer. On the other hand, Rho Ophiuchi looks bluer than the colour index suggests it should. Generally, however, the colours on the colour film correlate well with reality.

The fundamental limitations remain however, and shortly after the first deep-colour pictures were published by Miller alternative techniques for adding separate monochromatic exposures to make two- and three-colour separations appeared in the literature. The reconstruction of colour pictures from colour separations is not of course new: it is Maxwell's original idea; but it had not been tried in astronomy before.

Colour pictures from black and white plates

The use of the emulsions designed for the rather special requirements of astronomy immediately confers some important advantages on the prospective colour photographer. A most important benefit is the much greater efficiency of films and plates specially manufactured for astronomical use involving long exposures at low light levels. Reciprocity failure is minimised in the manufacture; and the speed of these materials can be greatly increased and reciprocity failure in some cases completely eliminated by the variety of hypersensitising techniques now routine at most observatories (see pp. 32–33). Another important advantage of these materials is their availability in a variety of spectral sensitisations, which enables almost any part of the spectrum to be isolated and used with other regions as a component of a multicolour reproduction. The making of colour separation negatives in no way interferes with more important astronomical practice, and in fact colour pictures can be made from plates taken at earlier times and for quite unconnected purposes. In some cases one or more suitable plates already exist in observatory archives awaiting combination into colour photographs. The picture of NGC 253 (Fig. 130) was made from plates taken on the AAT well before the three-colour project was thought of. Finally, of course, these plates are useful in their own right as calibrated records of objects in three or more colours with accurately defined passbands.

While the monochrome colour separations are made as a routine operation during the night (the more routine the better – thinking is notoriously unreliable in the early hours of the morning) the more complicated business of combining them into a colour picture can be tackled at a later time.

The first reported use of colour separation techniques in astronomy was by Merle F. Walker (1967), who superimposed plates obtained with the Lick Observatory 120-inch telescope in blue and infrared light and picked out red stars in the galaxy Messier 33. Walker, Blanco and Kuukel (1969) repeated this project using plates of the Magellanic Clouds obtained at Cerro Tololo Inter-American Observatory. They combined the plates using the Kodak dye-transfer process whose origins can be traced back to Edwards (1876) and Cros (1869). In its modern

TABLE 7. STARS NEAR ANTARES

Star	$B-V$	Star	$B-V$
White		*Light blue*	
π Sco	−0.19	ρ Sco	−0.20
δ	−0.12	ν	+0.04
β	−0.07		
σ	0.13	Mean	−0.08
τ	−0.25		
Mean	−0.10		
Blue		*Orange*	
48 Lib	−0.10	θ Lib	−1.02
λ	−0.01	α Sco	1.83
1 Sco	−0.05		
2	−0.07	Mean	1.43
13	−0.16		
d	+0.02		
N	−0.16	*Red*	
ω_1 Sco	−0.04	κ Lib	1.57
22	−0.11	ω_2 Sco	0.84
ρ Oph	0.24	19	0.84
ω	0.13	ψ Oph	1.01
χ	0.28	φ	0.92
24	−0.02	26	0.41
Mean	0.00	Mean	0.93

Stars are grouped according to their colour as photographed in Fig. 37(*a*). Within each group of stars are listed the stars' colour indices, taken from standard catalogues.

Fig. 38. *Central region of the Orion Nebula. In Joe Miller's dye-transfer picture the four Trapezium stars can be seen in the middle of the photograph, embedded in a green nebula which is the colour of the 4959/5007 Å pair of [O III] spectral lines. A ridge of Hα shows to the south. Contrast the colours with Fig. 85.*

In making this picture the dye-transfer process was adjusted to match the colour response of the dark-adapted eye (Fig. 18), which peaks near 5000Å. Thus the green [O III] lines were emphasised and dominate the picture. However, the idea in taking a photograph is to re-create an image true to reality, so that if the photograph and the nebula are compared they appear to the eye to have identical colours. If the green is emphasised in the photograph compared with the real nebula, it is doubly emphasised when viewed by the eye, and the result is unnaturally green.

form this process was (and is) capable of making very high quality colour prints (Tull, 1963). The original negatives are copied on to sensitised gelatin layers or mats. The material in the mats is altered by the exposure and subsequent development in such a way that its dye absorption properties vary according to the density of the originals. The mats are soaked in dyes of the subtractive primary colours (yellow, magenta and cyan) which are then transferred to an absorbent paper. Each separation negative has its own mat and the operation is undertaken sequentially with each mat in register on the absorbent base. This difficult operation was well established at Lick Observatory and in a well-illustrated article in *Scientific American* Joseph S. Miller (1976) described how dye-transfer colour photographs reveal much about the astrophysics of gaseous nebulae. These photographs (see Fig. 38) were made with plates taken on the venerable Crossley telescope, an instrument which owes its origins to Common's 3-foot instrument used for the original Orion photograph of 1883. This fascinating link with a pioneer of astronomical photography is fully explored in two recent articles by Remington P. Stone (1979).

Of the subtractive processes, the dye-transfer system offers the most flexibility in matching and balancing colour separations, but the process requires much skill and patience to achieve reproducible results and few examples of colour photographs of astronomical objects have appeared. The dye-transfer technique is by no means the only way of combining colour separation negatives. A method using individual dyed positives and the multiple printing of the negative originals on to colour negative films has been described by Alt, Brodkorb and Rusche (1974) and Rhim (1974).

Direct use of the negative original plates to produce colour separations in the complementary colours is an obvious and reasonable way to use the material. The subtractive technique, however, essentially simple as it is, is not ideal for obtaining the best results from the kind of photographic plates used in astronomy. These are often rather contrasty and have an unusually wide density range. It is often difficult to ensure that all three original plates have similar background densities and similar shapes to their characteristic curves, so that their contrasts are similar. The problem is mainly one of controlling the factors essential to good colour reproduction – contrast, curve shape and tonal range. They are not readily manipulated with the three-colour subtractive process operated in what is usually a darkroom specialising in monochrome photography. If the three original plates have different contrast, subtractive derivatives cannot be accurately colour balanced over the full density range. This is the problem of 'crossed curves' encountered earlier with conventional colour films. These variations in background density and the large tonal range and high contrast of the originals lead to burn-out of the highlights and difficulty in balancing the colour of the sky in the final prints.

Additive processes

Many of these problems can be avoided by reviving the long-neglected additive process, first used by Maxwell 120 years ago. This system has the enormous advantage of including a positive copying stage where the characteristics of the originals can be adjusted in a way which is not possible with the subtractive methods. Modern materials simplify the copying and subsequent colour printing stages to the point where it becomes a simple operation to make matched sets of colour separation positives from a wide range of original plates.

Combining the three positives into a colour picture can be accomplished in the manner of Maxwell, with three projectors. The combined images on the screen can then be photographed. The multiple projector technique has been used by Dufour and Goodding (1976) and by Dufour and Martins (1976) to make colour pictures of the Magellanic Clouds and of Centaurus A. In both cases the combined colour picture was recorded on conventional positive film or paper.

Alternatively the three images can be combined by sequential printing through red, green and blue filters onto a positive-working colour material. This technique has been used to make most of the colour pictures which appear in this book. A significant advantage of the technique is its simplicity and flexibility. There are few other colour processes where the user has (or indeed requires!) control over the contrast of the individual colours or where quite wide variations in exposure, density and contrast of the three original plates can be accommodated. The method can incorporate more advanced techniques of image manipulation such as unsharp masking or photographic amplification in order to improve the colour rendition, and we now describe some of these image manipulation techniques.

Unsharp masking

The negatives obtained from optimally exposed and processed astronomical plates are different from those found in normal photography, as shown in Fig. 39. These curves indicate the increase in blackening of the emulsion with increasing exposure. Blackening is measured by the transmission, T, of the emulsion: it is expressed in units of density, $D = \log(1/T)$. A density of 1 means that the film transmits 10% of the light incident upon it. Because a wide range of exposure is involved, the exposure axis is also marked on a logarithmic scale, where a change of one unit of exposure corresponds to a factor-of-ten change in that value. These curves reveal much of how a photographic material behaves on exposure to light and are therefore known as *characteristic curves* (or sometimes H and D curves after Hurter and Driffield (1890) who pioneered this form of presentation).

In Fig. 39 we compare the characteristic curves of an astronomical emulsion, Eastman Kodak type IIIaJ, and a commercially available camera-speed film, Eastman Kodak Plus-X. The exposure axis is relative – no attempt has been made to compare the speeds of these materials. The long, gently sloping curve of Plus-X is typical of a medium-speed low-contrast general purpose film. The characteristic curve is more or less linear over a log exposure range of about 3, corresponding to a subject brightness ratio of 1000:1. The film's response covers a large factor of relative exposure: it is tolerant to overexposure or underexposure. Any relative exposure in this range results in a density within the capability of normal photographic equipment and laboratories. The film does not produce such dark images that, in an enlarger, the faint light which struggles through the film in the negative-holder is swamped at the paper by light scattered in the enlarger's optics, nor is faint light from the image overwhelmed by leakage from the 'safe' light in the photographic 'dark'-room, nor does the faintness of the light falling on the paper demand time-consuming exposures to make prints. The maximum density obtained in Plus-X is about 1.6 above base and fog. The 1000:1 relative exposure in the scene being recorded is thus compressed into a dynamic range on the film of about 50:1. This compressed range is quite printable on a normal grade of photographic paper.

The high-contrast IIIaJ on the other hand has an extremely steep characteristic curve, the linear part of which covers a subject brightness range of only 10:1 but in doing so reaches a developed density in excess of 4. A quite short exposure of a bright object will rapidly produce the maximum density obtainable on this emulsion, somewhere in excess of 4 above the base fog level (the 10-minute exposure of the Eta Carinae Nebula in Fig. 28 is an example). A photographic density of 4 transmits only 0.01% of the light incident upon it, so the printing dynamic range is 10 000:1, quite unprintable on any type of photographic paper.

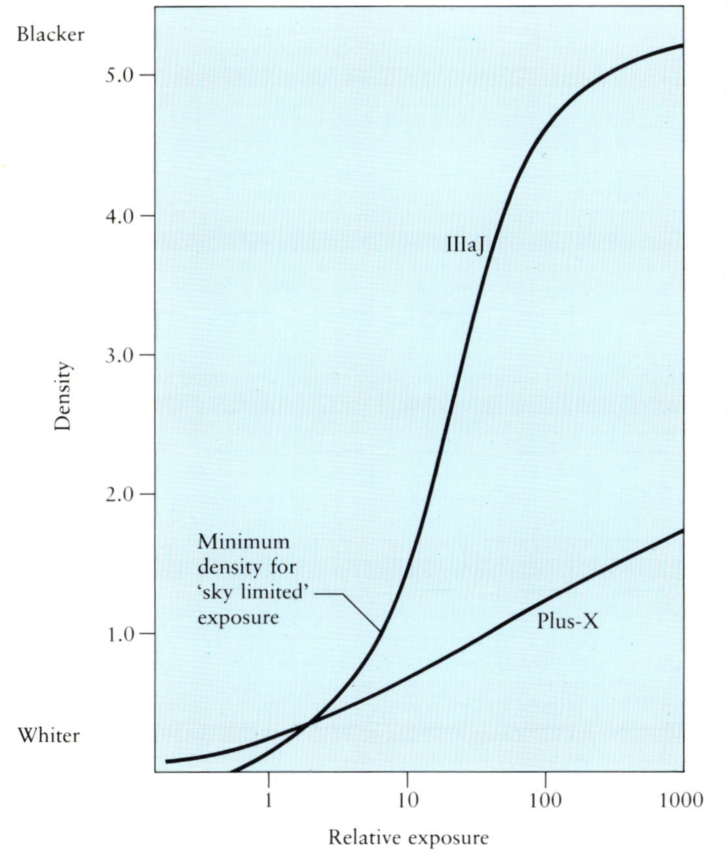

Fig. 39. *Characteristic curves. This figure compares a widely available camera film (Kodak Plus-X) with an astronomical emulsion (Kodak IIIaJ). The characteristic curve of each shows how increasing exposure to light causes an increase in blackening on the film. IIIaJ quickly reaches a much higher density than Plus-X (i.e. gets blacker), but Plus-X changes more gradually. Thus IIIaJ has a much stronger contrast than Plus-X. To show faint nebulae against the background light of the sky, a high contrast is necessary. On IIIaJ, short exposures have low contrasts, (the 'toe' of the curve). Maximum contrast occurs between densities 2 and 4, but at such high densities the picture gets more grainy. Thus to best detect faint nebulae a compromise sky background density of about 1 is used. A IIIaJ plate exposed to this density is described as* sky limited.

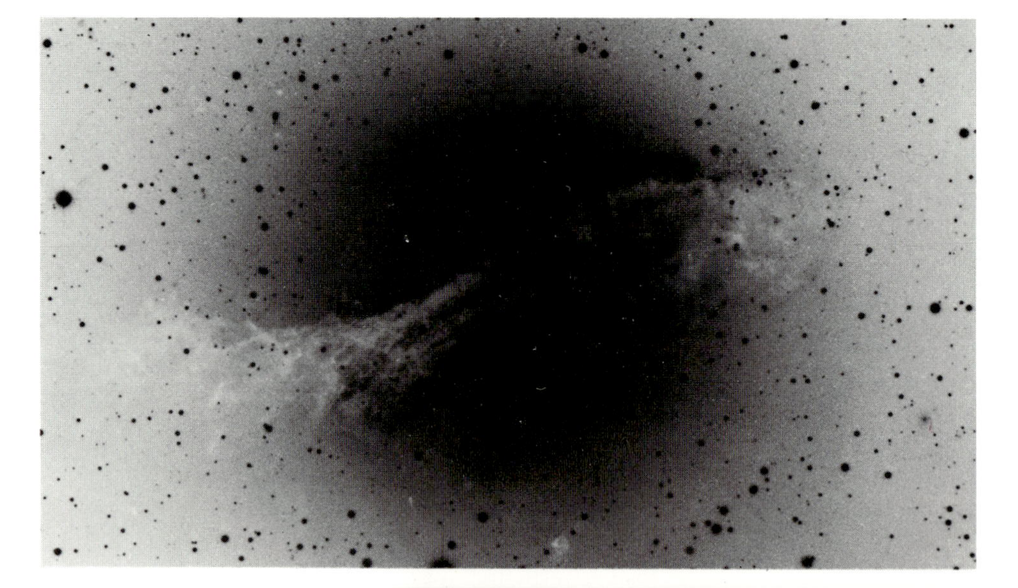

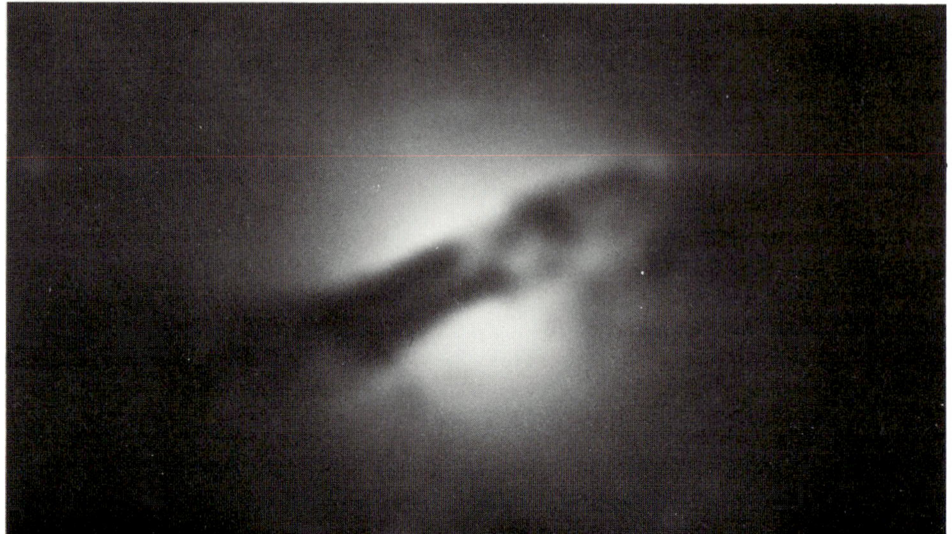

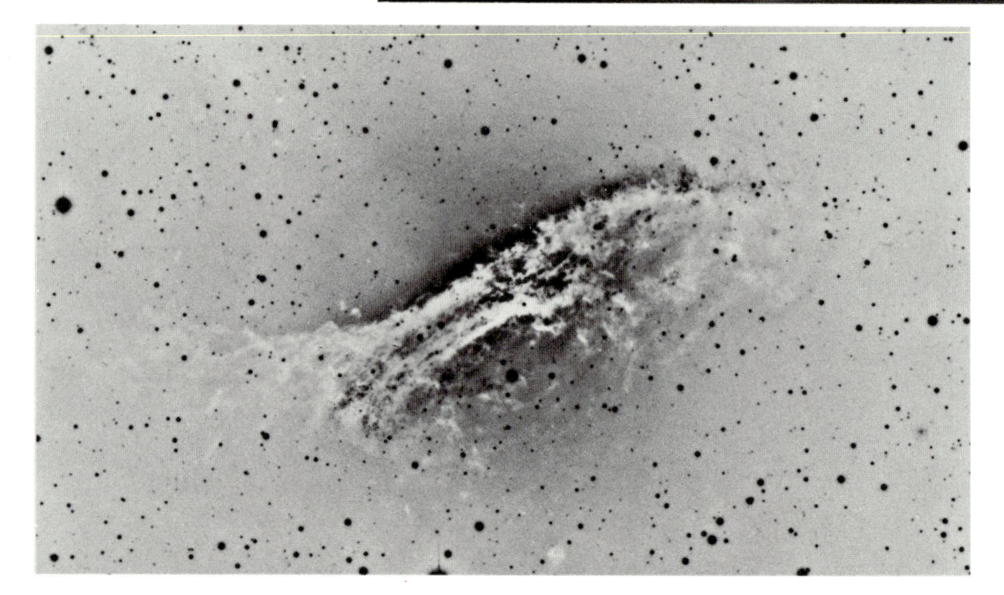

It is not surprising therefore that a technique called unsharp masking, should prove valuable in extracting usuable images from astronomical negatives.

As outlined by Nelson (1966), unsharp masking has been used for many years to equalise the characteristic curves of colour separations used in photomechanical printing processes and to emphasise fine detail in the resulting prints. The effect, although essential, is often small and the technique has been largely superceded by electronic methods.

An unsharp mask is made by making a positive contact copy of an original plate on a low-contrast film. The copy is blurred because it is made with the copy film in contact with the *back* of the original and because a diffuse light source is used. The glass acts as a spacer and the resulting positive contains only the blurred image from the original plate. When the processed and dried unsharp positive is re-placed in contact with the back of the original negative it effectively cancels most of the coarse structure leaving the fine detail either unchanged or enhanced. The sequence of events, described by Malin (1977), is shown in Fig. 40. Here, a deep AAT plate of the radio galaxy Centaurus A is reproduced (*top*) much as it appears to the eye. Fine detail in the dust lane is lost in the brightness of the relatively unstructured elliptical galaxy behind. An unsharp mask (*centre*) derived from this original plate carries little information about the dust lane or of the foreground stars but almost all the structure of the elliptical component is present. If this blurred positive is now combined with the sharp original negative and an emulsion-to-emulsion copy is made in the normal way, only the fine structure fails to cancel and this appears in the resulting positive, (*bottom*). The plate used for this exercise was the blue-light plate of Centaurus A from the three-colour image which appears in Fig. 133.

Another relatively nearby galaxy is an interesting subject for the unsharp masking technique, especially when the density of the mask is changed, as shown in Fig. 41. Here the prominent central bulge of the famous Sombrero galaxy, M104, has been progressively reduced in size to reveal more and more of the elusive structure in the disc of the galaxy. Other small objects on the plate, including some of the globular clusters known to surround M104, are unaffected by the process. By pictorially removing the bright central bulge it is possible to see through the galaxy to the disc beyond and pick out structures which are normally too faint to see on short exposures and hidden in long ones.

Fig. 40. *Unsharp masking in Centaurus A. From an original deep plate of Centaurus A,* (a) *(top), a blurred positive* (b) *(centre) is made by contact copying via a diffuser. The positive carries only the coarse detail present in the original and, when combined with it, cancels only the coarse part of the image, leaving* (c) *(bottom), the fine structure, to be printed from the combined mask and original. The process reveals detail otherwise hidden in the large, bright envelope of the galaxy.*

The image structure of any extended object can be explored by means of an unsharp mask, and the subject of Ainslee Common's famous picture, the Great Nebula in Orion, is an ideal example. The original plate used to make Fig. 42 was a 5-minute red light exposure made on the AAT during twilight – it does not need to be dark for photography of bright objects. In spite of the short exposure, the original was quite unprintable by any conventional enlarger because of the high central density. The region around the bright stars known as the Trapezium had a density of 3.80 and so transmitted only 0.16% of the light passing through the clear part of the plate. With a mask designed to reduce the dynamic range of the negative from 16 000:1 to about 30:1, a print with a remarkable amount of detail could be produced using a normal grade of paper. The Trapezium stars are just visible in the brightest nebulosity at the centre.

Photographic amplification

Unsharp masking is an easy-to-use contact copying procedure for exploring the overexposed parts of images. More often, however, astronomers are concerned with the other end of the exposure scale, where images of faint objects (signal) are barely visible above the developed density due to fog from the night sky airglow (noise). This is why astronomical emulsions are contrasty – to improve the output signal-to-noise ratio of the recording system. The apparent signal-to-noise ratio can be further improved by a quite simple contact copying process, first used by Malin (1978(*a*)), which requires the diffuse light copier mentioned earlier. Its success depends on the fact that non-saturated images tend to lie near the surface of the exposed (and developed) emulsion layer. This is due in part to light being strongly scattered as it enters the very turbid photographic layer. The effect is more pronounced in the finer-grained materials such as Eastman Kodak type IIIa than in the earlier generation of coarse-grained emulsions.

Copies of astronomical plates are usually made by contact printing the original onto the copy material with a distant point light source. This gives extremely high-resolution copies where images of all the silver grains in the original are transferred to the copy film. If, however, a diffuse light source is used, only the near-surface grains are imaged, but with their effective diameters much enlarged, whereas grains deeper in the emulsion produce shadows greatly diluted by the diffuse light and are too weak to appear on the copy. With a high-contrast copy film and careful control of exposure, an enormous improvement in the detection of low-contrast threshold images can be obtained. To complete the process the rather thin-looking positives are enlarged on to a hard grade of

Fig. 41. *Unsharp masking in the Sombrero. Internal details of the Sombrero Galaxy (M104) are hidden in this deep plate by the bright central bulge in a direct print from the original* (a) *(top). Unsharp masks with various characteristics can be prepared from the original to remove part of the central bulge and to reveal internal structure in the disc* (b) *(centre) or to emphasise the faint objects (globular clusters, distant galaxies) normally hidden amongst the bright light of the bulge* (c) *(bottom).*

bromide paper. For the best results the original plates must be of the highest quality, free from processing marks, emulsion defects, or optical vignetting. The technique has been extensively applied (Malin, 1981, 1982*a*) to the excellent plates from the UK Schmidt telescope and has led to the discovery of many new, low surface brightness objects. The technique's usefulness for astronomy was a serendipitous discovery during experiments to push the plates to their ultimate performance.

Amongst the most surprising discovery was a series of shells found by Malin and Carter (1980) to be apparently associated with otherwise quite normal elliptical galaxies. NGC 1344 was one of the first found and is the clearest example of this phenomenon (Fig. 43). The shells are about 56 kiloparsecs (kpc; 1 kpc = 3200 light years (l.y.)) from the nucleus and are remarkably sharp-edged. This property is quite distinctive and is present in many of the galaxies of this type discovered in a recent search of film copies of the atlas of the southern sky made on the UK Schmidt telescope. Several examples, including NGC 3923, also reveal a series of inner shells, much nearer the nucleus, when the plates are treated with the unsharp masking technique (Fig. 44).

Recent optical and infrared observations with the Anglo-Australian Telescope have revealed that the shells consist of stars very similar to those which make up the main body of the galaxy. The observations made by David Allen, David Carter and David Malin (1982) measure the most luminous part of the shell at only one thousandth the brightness of the night sky in the infrared; this is by far the faintest object for which infrared colours have been obtained. As always,

Fig. 42. *Orion nebula, masked. The delicate tracery of the Orion Nebula is revealed in this red-light plate. Although the exposure was very short (5 min) the central region was very dense on the negative and was printed through an unsharp mask to reveal the Trapezium stars in the brightest nebulosity. The general impression of outward flow indicates how the energy release in the central stars is causing the gas to compress in waves around them.*

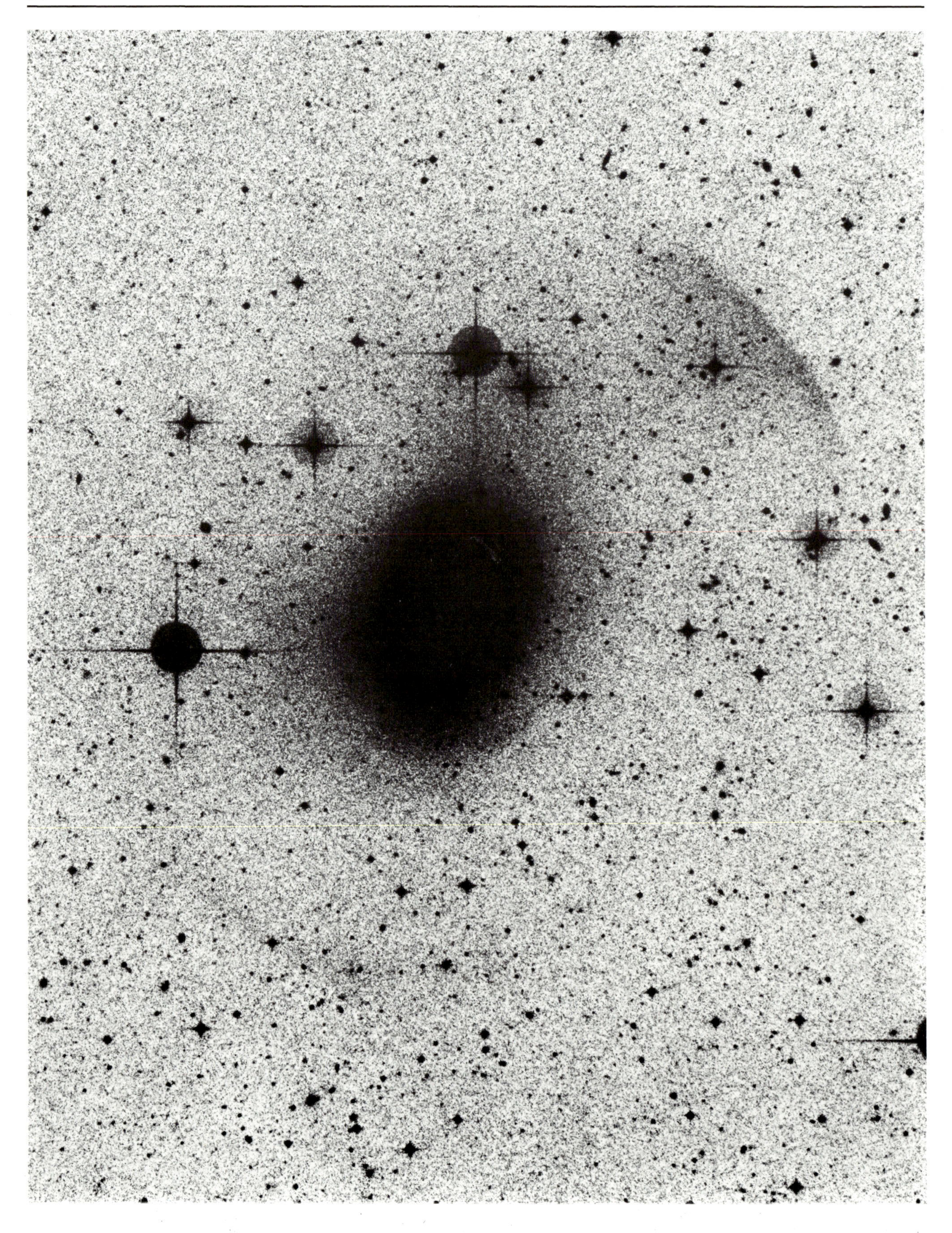

Fig. 43. *(left)* NGC 1344, *contrast enhanced. This is one of the best examples of what appears to be a new class of galaxy, ellipticals with shells. The brightest part of the shell (upper right) is barely visible on the original IIIaJ plate. Photographic amplification reveals its full extent and another, fainter shell at lower left. Faint features are more obvious against a light background, which is why very deep astronomical prints are reproduced as negatives.*

Fig. 44. *(below)* NGC 3923. *Many shell-type galaxies reveal unsuspected internal structure when photographic images are copied using an unsharp mask. In this example at least nine shells are visible alternating in their spacing away from the nucleus. It is believed that the shells are composed of stars displaced by an encounter with another galaxy, now merged into the giant elliptical.*

further observations on other galaxies of this type are needed to establish the generality of the observed properties but already several theorists have proposed mechanisms for shell formation in normal, isolated elliptical galaxies.

A galaxy which can in no way be considered isolated is NGC 4552, a conspicuous 11th-magnitude elliptical galaxy in the Virgo cluster. This ordinary-looking object, easily visible in a small telescope, had attracted little attention since Messier give it the number 89 in his famous catalogue, though it was known to be a rather weak and variable radio source. Photographic amplification by Malin (1979) of several deep plates from the UK Schmidt Telescope reveal this

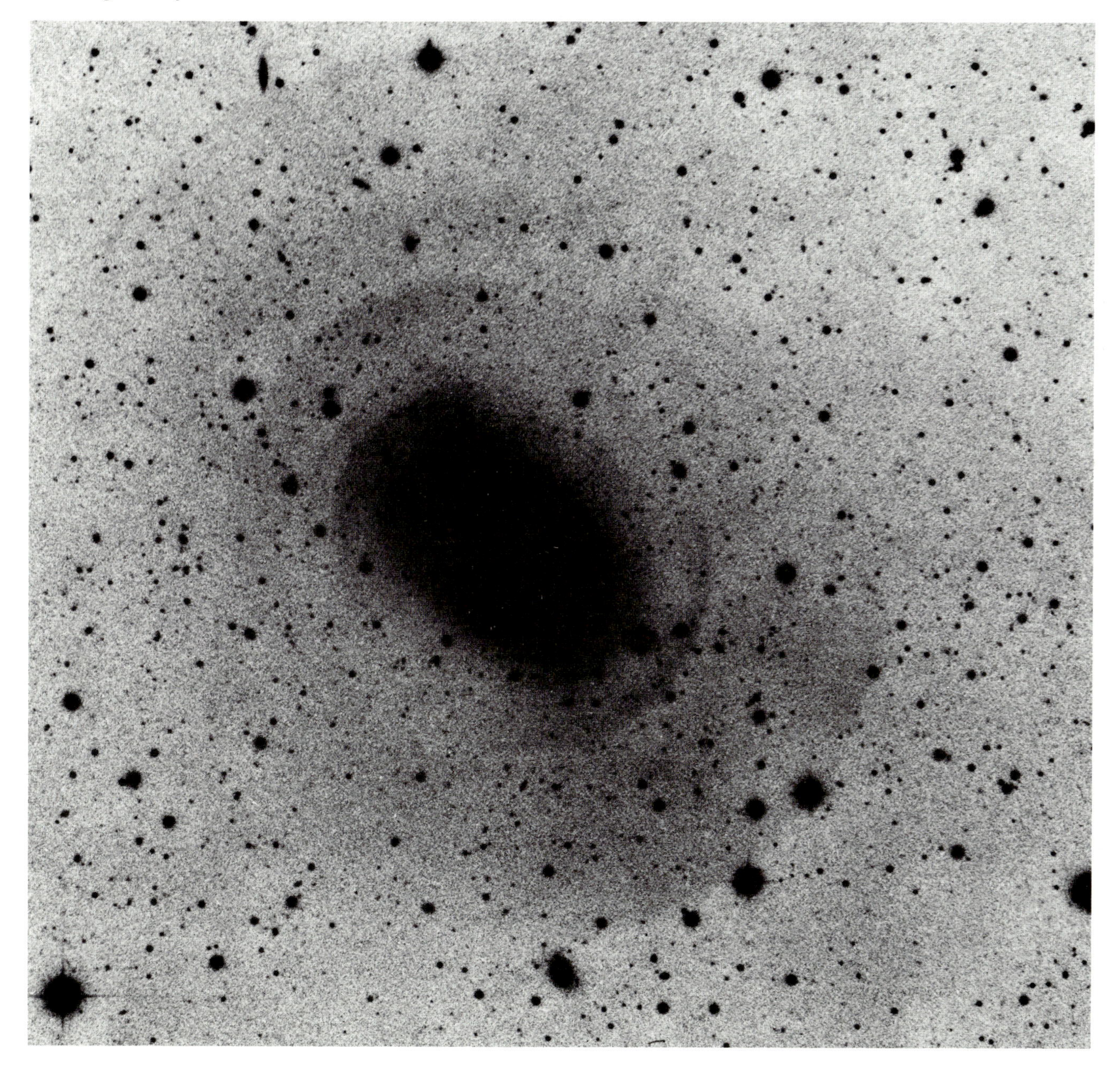

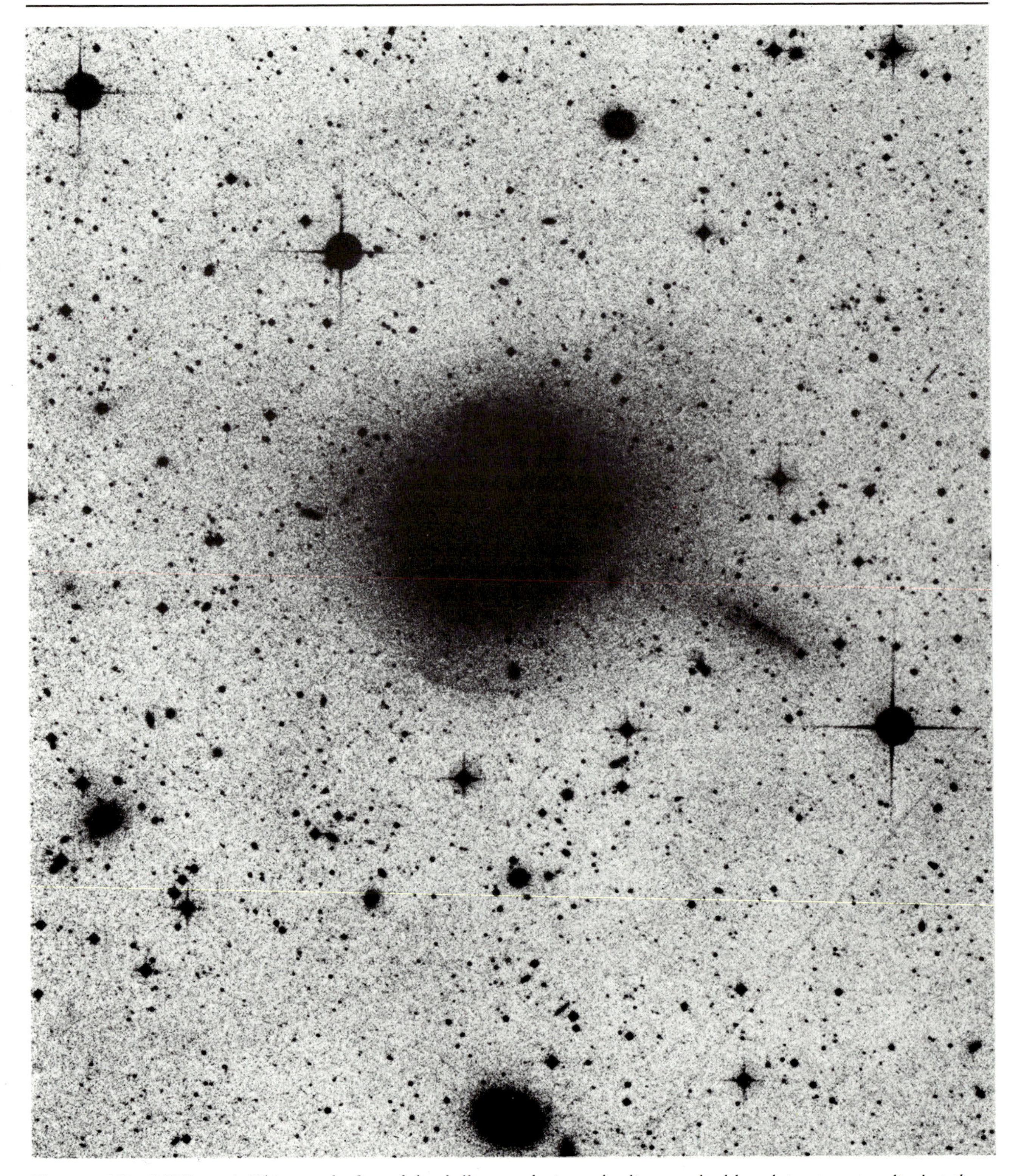

Fig. 45. *M89 (NGC 4552). This was the first of the shell-type galaxies to be discovered, although it was not realised at the time. The jet-like structure at right is probably all that remains of a companion galaxy, long since merged into the main body. The very faint shell (upper left) is probably the faintest astronomical object ever photographed and has a surface brightness much less than 1% of the night sky.*

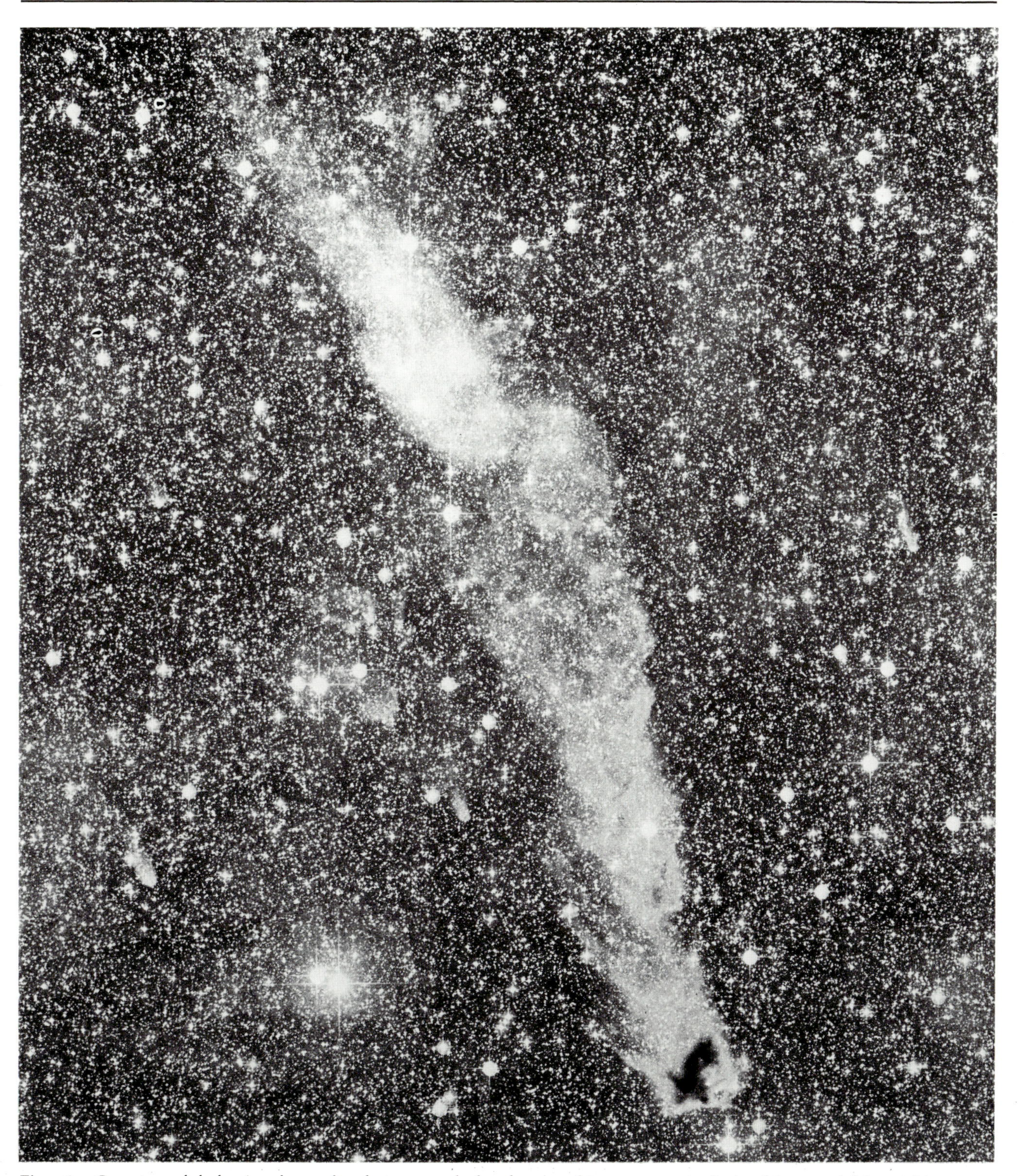

Fig. 46. *Cometary globule. As advanced techniques push the photographic process to its limits, fainter and fainter objects are revealed. The long tail of this cometary globule is illuminated by starlight, reflected and scattered by dust particles which have been swept by stellar winds from the dark nucleus at lower right.*

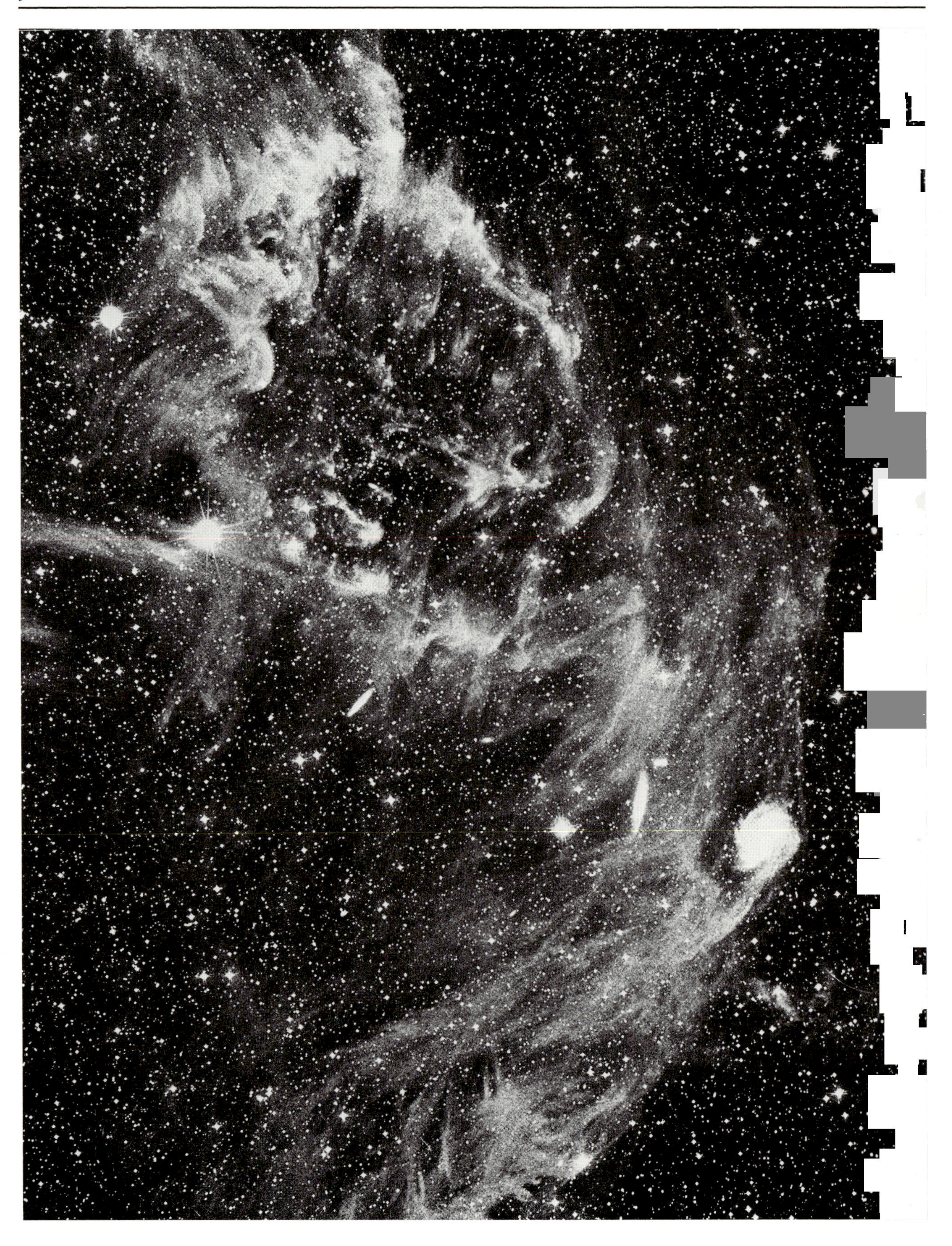

apparently normal galaxy to be something rather unusual, an elliptical galaxy with a jet extending 36 kpc (over 100 000 l.y.) from the nucleus, seen in Fig. 45. Closer inspection of the plates showed other peculiarities, namely some shell-like features visible near the envelope of the galaxy, just south and north-west of the nucleus, and a very large but very faint shell or arc of luminous material some 47 kpc (150 000 l.y.) north-east of the centre of the galaxy. Since its discovery Ken Elliott and David Malin have observed the brightest part of the jet spectroscopically on the AAT and found it to consist of stars, rather similar in colour to those comprising M89 itself. The most luminous part of the jet (burned out in Fig. 45) has a surface brightness of about 1% of the brightness of the night sky and we estimate that the faint north-east shell is 1.5 to 2 magnitudes fainter (about 0.2% as bright as the night sky).

We must not exclude our own Galaxy in this search for faint curiosities. The largest nebula so far found in our Galaxy is the Gum Nebula, discovered by Australian astronomer Colin Gum and including the southern constellations Puppis, Pyxis and most of Vela. In the direction of the Gum Nebula are found a collection of faint objects known from their appearance as cometary globules; more than 30 have been found, and one of the largest is shown in Fig. 46. Typically they have extremely opaque and compact heads with long, tenuous and faintly luminous tails, always pointing away from the centre of the Gum Nebula. In this central region are found two very bright stars Zeta Puppis and Gamma 2 Velorum, and the Vela supernova remnant and pulsar, any of which is capable of producing dusty tails from dark globules by means of radiation pressure. Some other faint nebulae are seen in Fig. 47. These nebulae are in the direction of the South Celestial Pole and are reflecting, not the light of a single nearby star as are the reflection nebulae in Chapter 7, but the faint integrated glow of our Galaxy. The surface brightness of these clouds, which are at a distance of 100–200 l.y., is only about 1% of that of the night sky.

Integration printing

The photographs made with the amplification process inevitably appear more grainy than copies made in the normal way because the effective area of the near-surface grains in the original is enhanced. With modest enlargements this is not intrusive and represents a real improvement in the visibility of the image. However, further gains in the detection of faint objects can be achieved, together with a reduction in grain noise, by printing in register the photographically amplified copies from additional plates of the same object. This has the effect of reinforcing the faint images appearing on each of the plates while reducing the apparent granular structure, which of course varies from plate to plate. This method is not new, having been described by Kohler and Howell (1963) and Richter (1975) amongst others, but it has not previously been applied to deep, photographically amplified positive derivatives made from original plates. The positives can be sandwiched together and enlarged as a single image, but a better method which can accept an infinite number of derivatives is to use the simple superimposition frame described in Appendix 1.

The effect of adding together derivatives from three plates of similar quality can be seen in Fig. 48. The subject is an interacting pair of galaxies in the Virgo cluster, NGC 4435 and 4438. Fig. 48(*a*) was made from an amplified derivative from a single UK Schmidt plate, enlarged (on the original) about nine times; (*b*) is the same field but made by superimposing images from three derivatives from separate plates. A marked reduction in granularity is obvious and many more faint objects can be seen. The 'plume' of faint material extending northwards from the interacting pair has not previously been detected. The satellite trail (noise, in this context) which runs across the bottom of Fig. 48(*a*) has almost disappeared in 48(*b*) when combined with images from plates without that spurious image.

All this darkroom wizardry is applicable to photographic materials from any source, not necessarily optical telescopes. It can be used on normal but abused pictorial negatives (Malin, 1982*b*). It represents only a fraction of the possible ways in which photographic images can be manipulated, and no mention here has been made of the sophisticated digital techniques available to those with sufficient computer power.

Illusion or reality?

How accurate are the colours appearing in the photographs produced by adding colour separation negatives manipulated by these methods? Unfortunately there is no ready answer. The five fundamental types of colour reproduction discussed by Hunt (1975) all depend on a visual match of the original scene with the coloured reproduction. Only Hunt's 'corresponding colour reproduction' allows for the viewing and taking conditions (in terms of the type of illuminant and luminace level) to be different. Even here it is assumed that both original and reproduction

Fig. 47. *South Celestial Pole. The extremely patchy distribution of dust in the direction of the South Celestial Pole is shown in this contrast-enhanced image of a UK Schmidt plate. As well as reflecting the feeble light from our Galaxy, these clouds obscure and redden the light of stars and galaxies beyond. Not all such clouds reflect enough light to be visible and so their existence can only be inferred from colour measurements of more distant objects.*

are examined at high light levels where the eye can distinguish the full gamut of colours. As we saw in Chapter 1, the night sky, apart from a few stars and some planets, is generally visualised as a colourless place; the eye requires relatively high light levels for colour discrimination. It is therefore not possible to match the reproduced colours of astronomical objects with their originals.

The situation is further complicated by the nature of the objects themselves. Gaseous nebulae emit most of their visible radiation in the form of narrow emission lines. Hunt does not deal with this (terrestrially rare) form of subject matter in his discussion on colour reproduction. It is known from rules formulated by Grassman (1853) that the appearance of colour is independent of its physical origin for the normal observer and it has been shown by Maxwell (1860), Guild (1931) and Wright (1928) that the sensation of all colours in one's mind can be reproduced by mixing either three monochromatic or three broad-band filtered lights in various proportions. (There are some practical limitations to this concept, mainly concerned with small negative quantities involved in the colour response curves but these can be ignored in the present argument.) Thus it is possible to reproduce the visual appearance of, for example, a monochromatic sodium street light by mixing red and green coloured lights in suitable proportions. The perceived accuracy of reproduction therefore depends more on the proportion of the coloured lights than their actual colours.

The relative proportions of colours in additive astronomical photographs are determined by the spectral sensitivity of the plates and their associated filter passbands. These in turn have been chosen as far as practicable to cover the visible spectrum uniformly.

Fig. 48. *Integration printing: a way of adding two or more identical images on separate plates. Since the wanted signal, the stars and nebulae, are common to all the plates and the grain structure (noise) is different for each, combining many plates enhances the one and reduces the other. Here, the result from a single plate* (a) *(left) is compared with the combined image* (b) *(right) from three plates of the same object, the galaxies NGC 4435–38 in the Virgo cluster.*

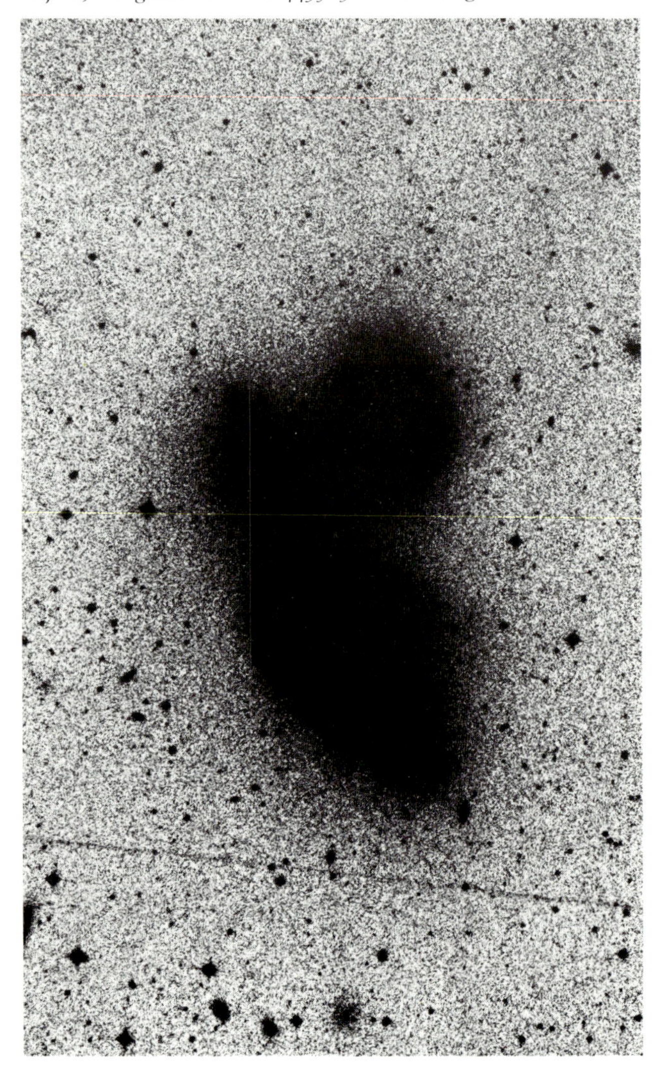

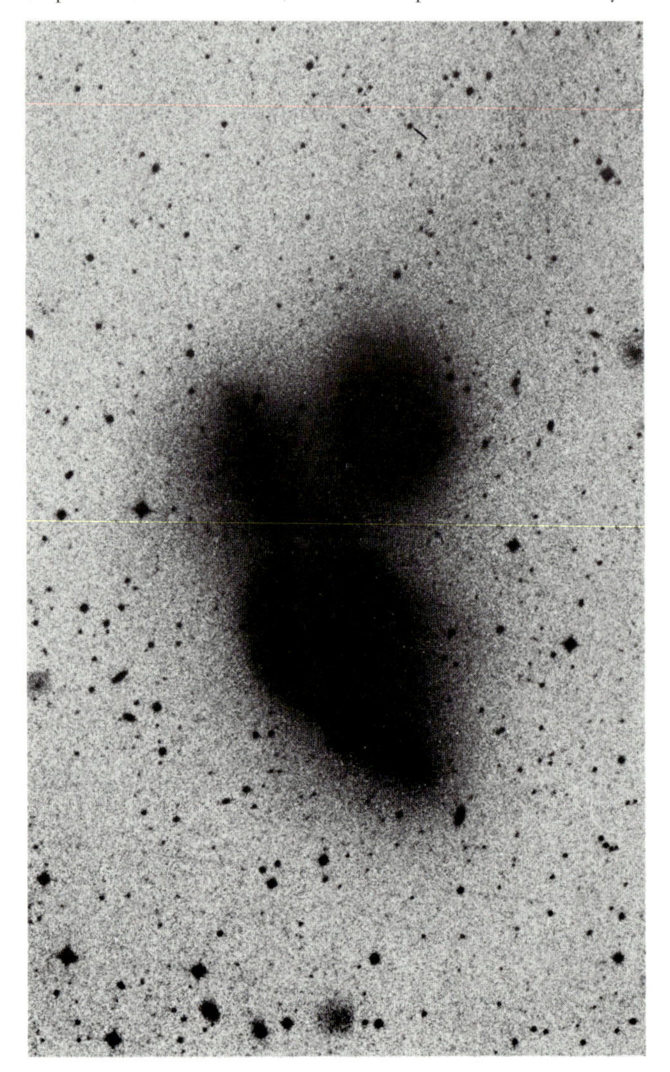

The relative proportions of light transmitted by positives made from these plates are determined by reference to a step wedge which appears on each plate and whose colour *is* accurately known. Thus colour balance is correct with reference to a secondary standard, and it is assumed that the astronomical images are likewise combined in their correct proportions. This is not to say that the colour rendering is in any sense 'accurate'. A more appropriate term is 'representative', in that each spectral band is given more or less equal weight. There are no significant gaps in the the spectral response but both luminance and contrast have been increased by an arbitrary amount to yield a usable colour image.

Finally, it is necessary to produce some permanent combined image from the three separate positives. While coloured lights can be mixed to give pure spectral hues, the coloured dyes used or created in photographic print materials can not. The ultimate limitation to colour accuracy is set by the photographic products used to superimpose and record the coloured image.

In the future we can look to Lippmann's (1891, 1894) method for giving spectrally correct colour rendition by reflection viewing. He used an extremely fine-grained, almost transparent photographic emulsion (of a type which still bears his name) which was exposed to an image through the supporting base. The surface of the emulsion was supported on a bath of mercury. Light passing through the emulsion was reflected by the mercury surface and gave rise to standing waves by interference in the emulsion layer. The image of the interferogram was photographically processed. The positions of the standing waves throughout the thickness of the emulsion layer were then defined by developed silver. By viewing the image so formed in reflected light the original, spectrally correct colours were reproduced. The extreme slowness, limited viewing angle and generally inconvenient nature of the technique have precluded its practical application. However, a recent paper by Lindegren and Dravins (1978) proposes Lippmann's system as a means of recording stellar spectra where the spectral information is recorded as standing waves within the emulsion layer to be retrieved – quantitatively – by means of a micro-spectrophotometer. The difficulties are considerable but the possibilities endless, and not only for astronomical colour photography. Until the time when this method becomes practical the astronomical colour pictures produced by the additive method are as accurate as any, and in the rest of this book we describe what they show.

Ainslee Common. Victorian pioneer of astronomical photography, Common took the first pictures which showed stars which had not been seen before. His pictures of the Orion Nebula, obtained 100 years ago, marked the beginning of astrophotography.

Orion Nebula. Royal Greenwich Observatory photographer D. A. Calvert used the three-colour addition method of colour astrophotography to make this photograph in 1977. The originals were taken by Peter Andrews with the Isaac Newton Telescope in Sussex before it was moved to La Palma in the Canary Islands.

Chapter Three

Dust clouds of Sagittarius

The Milky Way

From the time he could formulate the thought, Man must have wondered about the Milky Way. Today, the knowledge that when we look at the Milky Way we are looking at a galaxy from within a spiral arm can only increase our wonder; but such knowledge was not easily won and only within the last 60 years or so has the true nature become apparent of the magnificent stellar system of which we are a part.

Our Sun is one star in a collection of more than 10 billion called The Galaxy. This is a flattened disc in shape, somewhat like a fried egg. The Galaxy is, let us say, 100 000 l.y. in diameter and 3000 l.y. thick. Our Sun is in no particular select place, and lies roughly half-way out from the centre (25 000–30 000 l.y.) and centrally placed within the thickness of the disc. We view our Galaxy from the only position we ever can, near an insignificant star in a spiral arm (Fig. 49). When we look out from our planet we see the naked-eye stars, our immediate neighbours in space, mostly within a few hundred light years of the Sun. Beyond lies the rest of the Galaxy. Through the sprinkling of local stars, our Galaxy is only apparent when our line of sight passes through its greatest thickness in the plane of the disc. There we see the faint haze of the Milky Way crossing the sky. At right angles to the plane of the Galaxy the concentration of nearby stars hardly changes, but no longer do we see the misty veil of multitudes of stars too distant to be distinguished.

The Milky Way appears brighter when we look towards the centre of the disc than when we look away from the centre, because we view the most populated part of the Galaxy. The effect is not as dramatic as one might hope, because our view is obscured by dust which lies between the stars. There is not much dust in space – on average one speck of dust in a cathedral-sized volume (very clean by household standards!) – but the distances between stars are so large that the cumulative effect of these rarefied dust grains is noticeable. In fact the concentration of dust into a thin layer is so marked that over much of its length the Milky Way appears to be split into two irregular but roughly parallel bands of luminosity.

The centre of our Galaxy lies towards the constellation of Sagittarius. The wide-angle photograph of this region (Fig. 50) shows clearly that a huge part of the central bulge of our Galaxy (the yolk of the fried egg) is obscured by a lane of dust whose density and distribution vary enormously. In some parts the distant stars appear dimmer as if seen through a smoky haze; elsewhere they are completely obscured.

At first it was not clear that dust was the reason why there were patches of sky with few visible stars. In 1784, Sir William Herschel, the famous discoverer of Uranus, was 'star gaging' in the constellation of Ophiuchus which adjoins Sagittarius in the southern Milky Way. As he scanned the region now identified with the Rho Ophiuchi dark cloud, he was heard to exclaim 'Hier ist wahrhaftig ein Loch im Himmel!' (Here is surely a hole in the heavens!).

This statement effectively encapsulates Herschel's thoughts about the patchy distribution of stars in the plane of the Galaxy. He believed that the dark regions were devoid of stars and that he was able to see through the Milky Way to the void beyond. He believed, too, that the starry clouds were condensing or fragmenting into 'clustering collections', a view

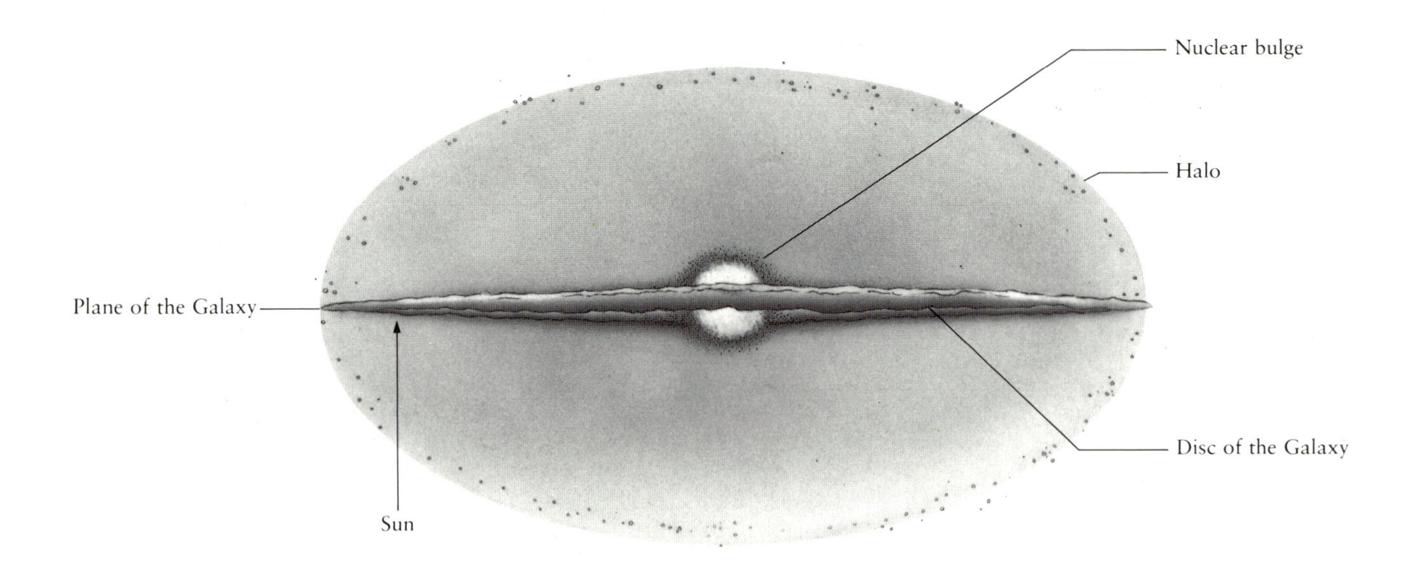

supported by the centrally condensed nature of other objects he had seen, such as globular and open clusters of stars and even nebulae far beyond our Galaxy. Although Herschel had no way of knowing that his tiny nebulae were star systems or galaxies in their own right, he did note that they showed a marked tendency to avoid the plane of the Milky Way. The alternative explanation, that all these effects were due to interstellar dust in clouds congregating near the Milky Way, had to await the fruitful partnership of photography with astronomy which began in earnest 100 years later.

Dust amongst the stars

One of the first to exploit the speed of the newly introduced dry gelatin plates was the famous American observer E.E. Barnard, who had an advantage rare in astronomers before or since – he had earned his living as a photographer's assistant! In 1884 Barnard began to take long-exposure photographs of the Milky Way with what he described as a 'portrait lens', which seems to have been a 31-inch focal length objective working at about f/5 and covering 8 × 10 inch plates.

In his early publications, which contain some of the first pictures of extra-galactic objects as well as wide-angle views of the Milky Way, Barnard (1890 *a*) was clearly excited about his photographs. To him they

> *showed for the first time in all their delicacy and beauty the vast and wonderful cloud forms, with their remarkable structure of lanes, holes, black gaps and sprays of stars as no eye or telescope can ever hope to see them.*

His enthusiasm did not affect his judgement however.

Fig. 49. *(above) Section through our Galaxy. The Sun is offset to one side of the galactic centre but in the central plane of the Galaxy. We view many stars when we look in the disc of the Galaxy and fewer when we look out of the Galaxy through the galactic halo. The Milky Way is our view of the galactic plane.*

Fig. 50. *(right) The Milky Way in Sagittarius. This wide-angle photograph was made with a conventional camera attached to the top end of the AAT while the telescope itself was being used to photograph M16, the small red patch near the centre of this picture. The brightest region is the central bulge of our own Galaxy, partly obscured by the dust lane crossing it.*

Only gradually did he come to believe that the dark lanes in the Milky Way were due to intervening matter. Barnard (1919) tells how a chance sighting of terrestrial clouds made him believe it:

> *All that is needed to make these dark bodies visible is a luminous region behind them. This is supplied in one way by the rich stellar regions of the Milky Way. An excellent example of how such a thing may be possible is shown by a phenomenon that presented itself to me one beautiful, transparent, moonless night in the summer of 1913, while I was photographing the southern Milky Way with the Bruce telescope. I was struck with the presence of a group of tiny cumulous clouds scattered over the rich star clouds of Sagittarius. They were remarkable for their smallness and definite outlines – some not being larger than the moon. Against the bright background they appeared as conspicuous and black as drops of ink. They were in every way like the black spots shown on photographs of*

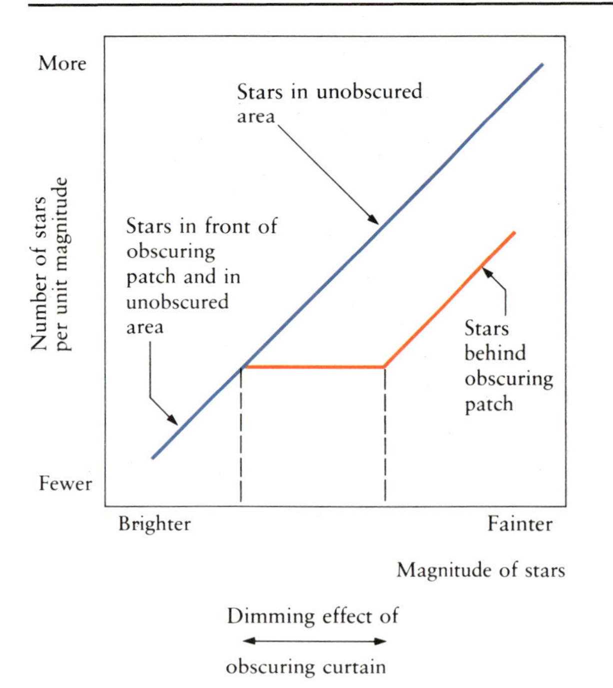

Fig. 51. *Wolf diagram. Wolf compared star counts in an obscured area with counts in a region nearby and discovered the dimming effect of the curtain of obscuration on the stars behind it.*

> *the Milky Way, some of which I was at that moment photographing. The phenomenon was impressive and full of suggestions. One could not resist the impression that many of the black spots in the Milky Way are due to a cause similar to that of the small, black clouds mentioned above – that is, to more or less opaque masses between us and the Milky Way. I have never before seen this peculiarity so strongly marked from clouds at night, because the clouds have always been too large to produce the effect.*

By 1919 his impressions had been transformed into certainty (Barnard, 1919):

> *I did not at first believe in these dark obscuring masses. The proof was not conclusive. The increase of evidence, however, from my own photographs convinced me later, especially after investigating some of them visually, that many of these markings*

Fig. 52. *Milky Way. This representation of the whole Milky Way was painted from a large number of photographs. The superimposed rectangle is the area of the enlarged section in Fig. 53. The Milky Way band stretched from left to right is split by a thinner band of dust marking the galactic plane. The Milky Way is broader and brighter in the centre of the picture towards the galactic centre, and there is more dust showing as large patches up to 20° in diameter (the area of four rectangles in the co-ordinate grid).*

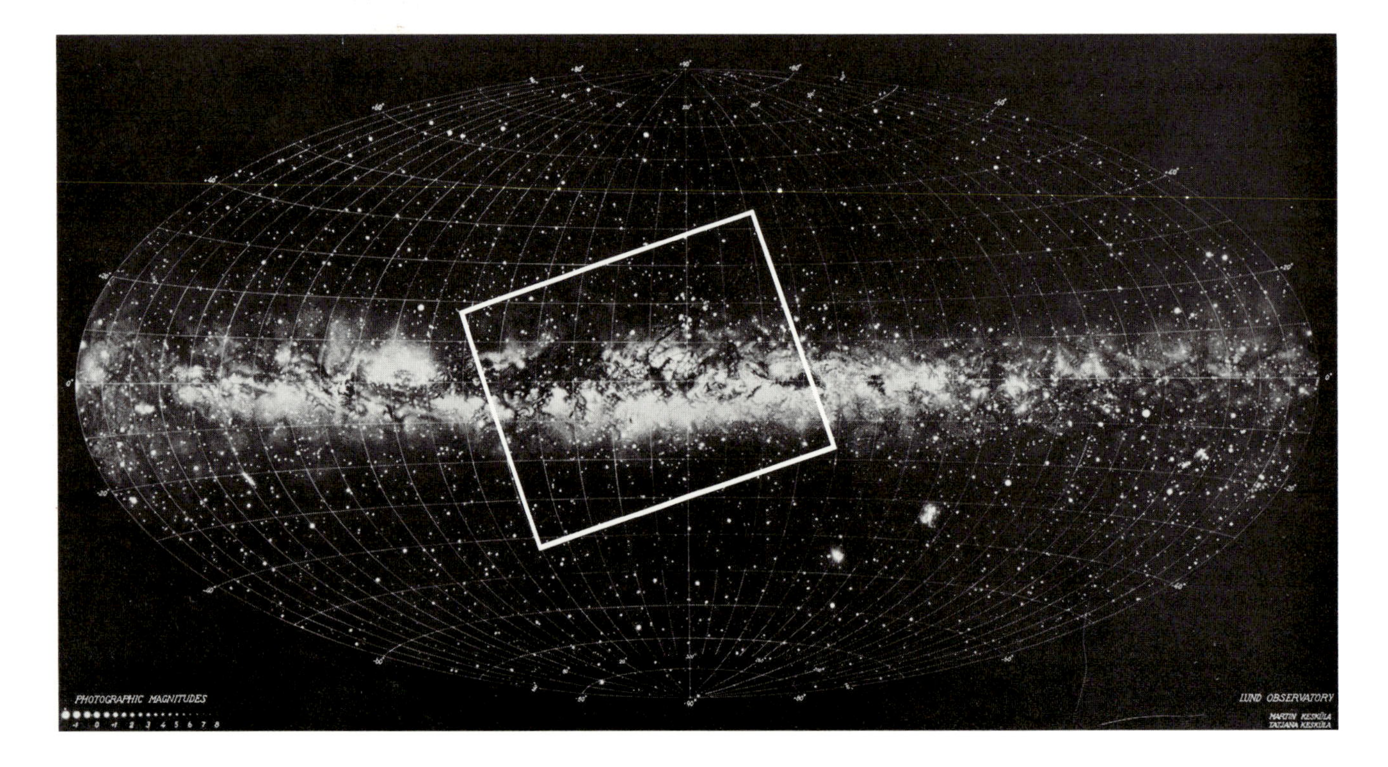

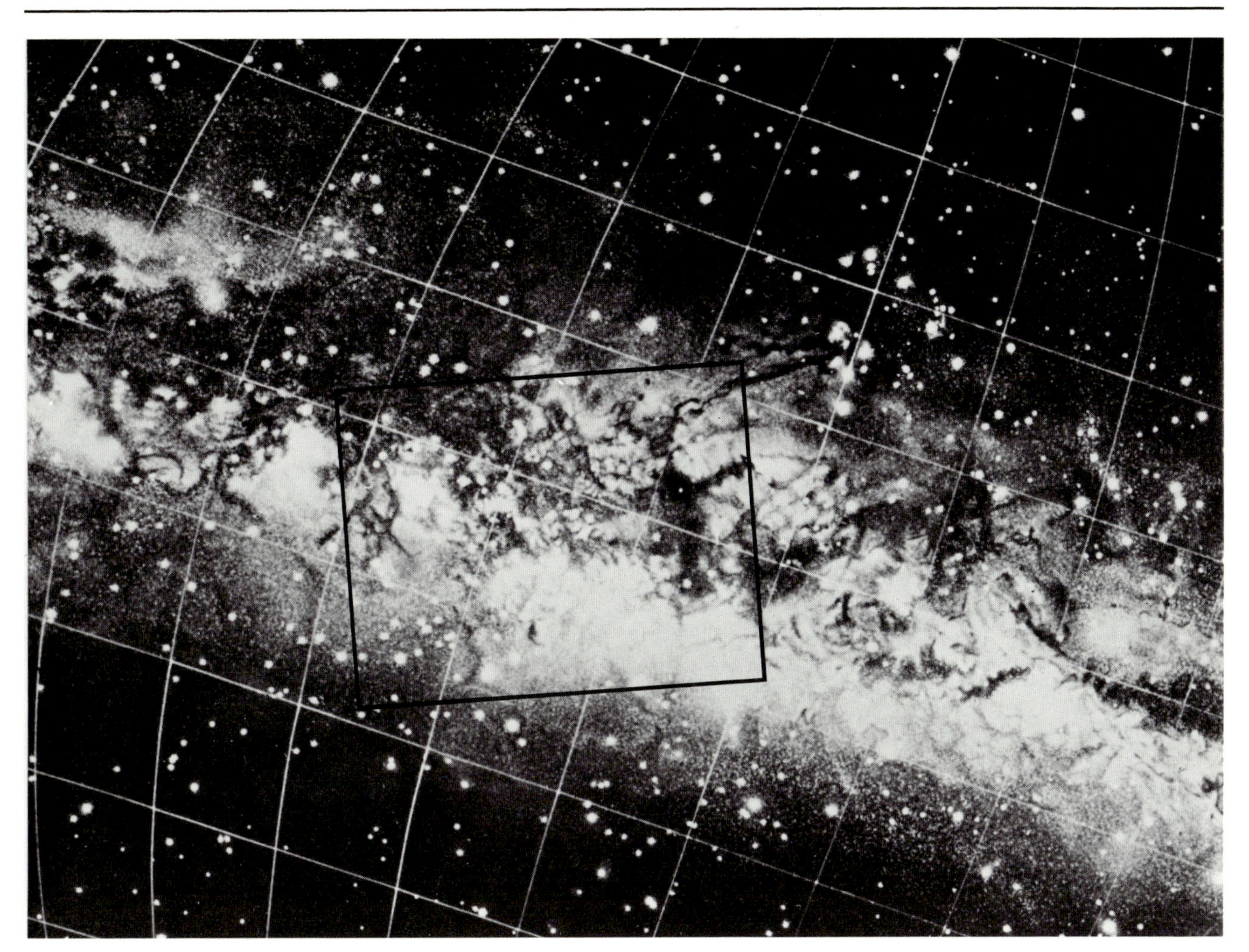

Fig. 53. *Galactic centre region. This enlargement from Fig. 52 shows the Sagittarius region. Compare with the colour photo in Fig. 50. Dust clouds up to 10° across are prominent, some arranged in long streaks, or twisted ropes.*

> *were not simply due to an actual want of stars, but were really obscuring bodies nearer to us than the distant stars. In this way it has fallen to my lot to prove this fact. I think that there is sufficient proof now to make this certain.*

In spite of Barnard's by now firm belief that the dark clouds were not holes torn in the starry background, doubts lingered on in some minds – but not for long. Using material from wide-angle photography as Barnard had done, Max Wolf of Heidelberg was able in 1923 to count stars in adjoining obscured and clearer regions. He concluded that the dark areas were clouds of dust, because the star counts in the obscured areas mimicked the star counts in the clearer areas, with an overall dimming as if the stars were seen through an obscuring curtain (Fig. 51). The final proof came from the American astronomer Robert J. Trumpler in 1930. He studied the way that the angular size of clusters of stars correlated with the stars' apparent luminosity. On the admittedly gross assumption that all clusters are the same size, their angular size shows their distance (the smaller, the further). Their constituent stars' luminosity should dim in a way correlated with small angular size. Trumpler found, however, that there was an extra dimming of the more distant clusters. The simplest explanation was that the apparent luminosity–angular size relation was disturbed by some intervening absorbing material. Trumpler's work was supported by similar conclusions drawn from the distribution of extra-galactic nebulae – a point that Herschel had missed 150 years before.

The work of Trumpler and his contemporaries gave self-consistent picture of an all-pervasive, non-luminous interstellar medium responsible for the progressive dimming of stars and star clusters with increasing distance from the Sun, an effect which becomes progressively more noticeable towards the plane of the Galaxy. But the distribution of the obscuration is not uniform and the size and thickness of the opaque patches seems to vary in an apparently

however still possible to visualise the cloud as a hole in the patchy distribution of stars.

The next photograph (Fig. 55) in the series was made in 1905 by Barnard and has been copied from his *Atlas of the Milky Way* (1927). The picture is centred on a part of the constellation of Sagittarius which had intrigued him 15 years earlier when he began to photograph the sky with his wide-angle camera. Of the plate centred on the region which he took in 1889, Barnard (1890 *b*) says:

> *This last plate shows an object that I wish to call attention to. For many years I have observed in my comet seeking a most remarkable small inky black hole in a crowded part of the Milky Way. This singular object, with one or two faint stars in its following part, is in about* $\alpha = 17^h\ 56^m$, $\delta = -27°\ 50'$. *I have not seen this black hole mentioned anywhere. It is about 2' in diameter, slightly triangular, with a bright orange star on its n*[*orth*] *p*[*receding*], [*i.e. north-western*] *border and a beautiful little cluster following. There are other dark holes and vast gaps near this, but nothing so remarkable in the entire circuit of the Milky Way . . . The exceeding beauty of a glass positive from this plate is beyond description.*

Later the 'inky black hole' was listed by Barnard as No.86 in his *Atlas of the Milky Way* (Barnard, 1927) and the cluster identified as NGC 6520. It can just be seen at the centre of Fig. 55.

Though Barnard was well aware that the luminosity surrounding the region was due to multitudes of stars, his photographs fail to distinguish or *resolve* most of them as individuals because their density is far too high. A much larger plate scale and fine-grained emulsion are needed to separate them clearly. Fig. 56 is a modern view of Barnard 86, made on just such an emulsion with the Anglo-Australian Telescope.

Finally in this series we show a colour photograph (Fig. 57), a picture which we feel would have given Barnard himself much satisfaction. The colour picture reveals a common characteristic of many of the larger dark clouds; they have a centre which seems completely opaque but which gradually

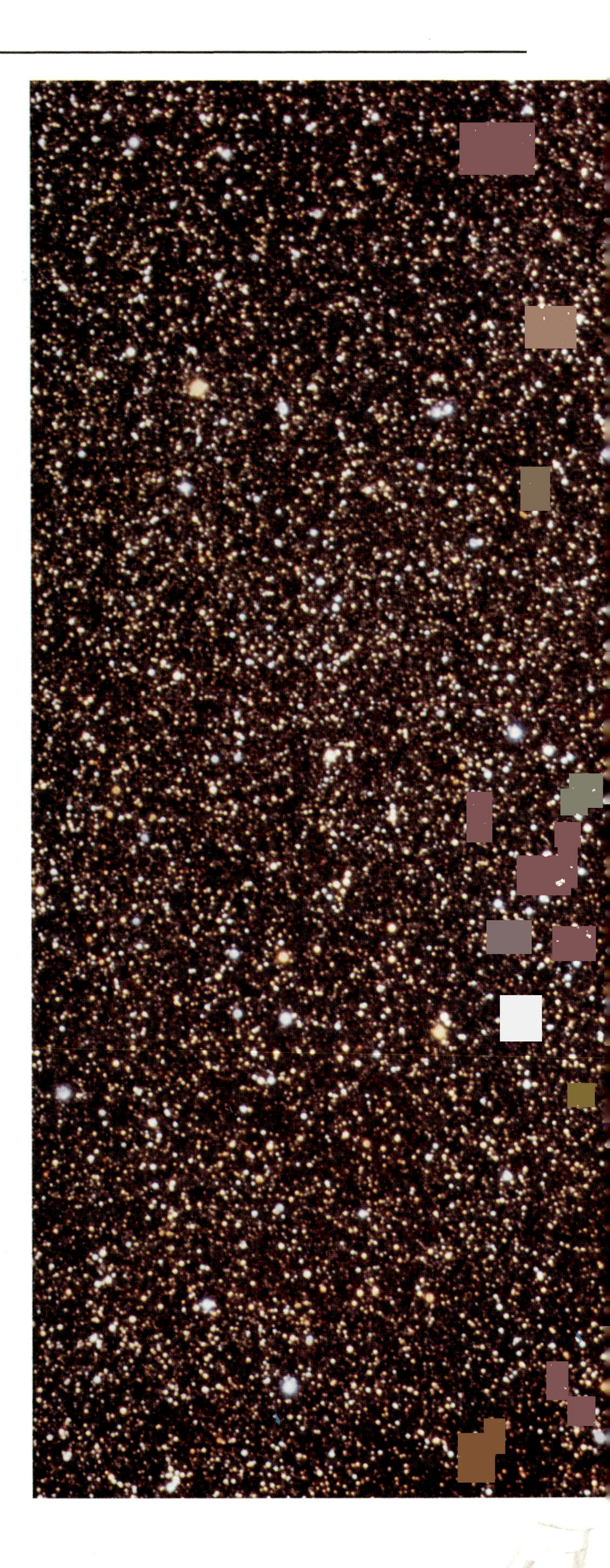

Fig. 57. *Barnard 86. As the preceding pictures have revealed, the dark cloud is set against one of the most populous regions of the Milky Way, and the multitudes of distant stars give the background a yellowish colour. This background makes an excellent contrast with the bright blue stars of the open cluster NGC 6520. In the cluster itself are two yellow giant stars, with another yellow star projected near the edge of the dark cloud. Parts of the cloud have very sharp edges, with dense material, but other parts have more transparent edges through which dim Milky Way stars can be seen. These dimmer stars behind the cloud are reddened.*

thins towards the edge. In part this is because the cloud is denser in its centre, and in part because light from stars seen through the centre has to travel through a longer path of dust than light from stars seen through the edge.

Interstellar reddening

Interstellar dust has an obscuring effect which is evident from the dimming of the stars behind it, but it also alters the colours of the stars, making the starlight redder. This is called interstellar reddening. It can be seen in Barnard 86 (Fig. 57) where the dimmer stars seen through the edges of the cloud are redder than the generally yellow brighter stars of the star clouds (the so-called *field stars*). The colour of the field stars is partly due to the nature of the stars themselves (they are predominantly yellow giants) and partly because a veil of interstellar dust covers the whole area of the photograph – Barnard 86 is a particularly strong and compact obscuring cloud. The brightest stars of all, the relatively undimmed stars of the foreground galactic cluster, are predominantly blue. The dimmer stars are fainter because they are more distant and/or more obscured, and Fig. 57 shows that there is a strong correlation between the amount by which dust obscures starlight and the amount by which it reddens it.

The colour picture of Sagittarius (Fig. 50) also shows the effect clearly. The picture is cut in half from left to right by a dark band which contains pink emission nebulae (Chapter 6), particularly the Lagoon Nebula, M8, right of centre. The band represents the centreline of the Milky Way (the galactic equator). Above and below the galactic equator lie the brilliant star clouds of Sagittarius, Scorpius and Ophiuchus. Silhouetted against them are dark lanes of opaque dust; particularly prominent are the dark clouds of Ophiuchus (mid right). Around these dark clouds the

Fig. 58. *Red sunset. An especially fierce outbreak of distant bushfires in central Australia had injected large amounts of dust into the atmosphere when this photograph was obtained from Siding Spring Mountain in 1974. Light from the Sun passing through the dust was thus very reddened and produced dramatic sunsets.*

Fig. 59. *Setting star trails. Stars set over the AAT dome in this time exposure. The star trails have various colours due to the individual colours of the stars, but in each case the star trail gets fainter and redder as the star dips to the horizon. Dust particles in the atmosphere, and even the air molecules themselves, scatter starlight and redden it.*

stars of the Milky Way have an orange tint, which contrasts with the whiter stars at the boundary of the Milky Way well above and well below the galactic equator. The effect of reddening is marked near the brightest patch of the Milky Way (below and to the right of the Lagoon Nebula). The upper boundary of the bright patch has a yellow brown hue, because this boundary is caused by the central obscuration of the Milky Way, whereas the lower boundary is white, or even blue, because this boundary is free of obscuration and is caused by an actual lack of stars.

It is clear that interstellar obscuration and interstellar reddening are correlated. What is the reason for this?

It happens that grains of interstellar dust are of about the same size (about 1 μm) as the wavelength of light. Light waves from the stars of the Milky Way pass into the dust and interact with it – some of the light is redirected (scattered) and some is converted to heat (absorbed). Fewer of the light waves pass on through the cloud. The efficiency of these scattering and absorption processes depends upon the ratio of the size of the particle to the wavelength of the light. Simply put, particles which are large compared with the wavelength are more effective scatterers than are smaller particles.

You can prove this by a bath-time experiment. Sit in the bath with knees protruding through the surface of the water. These make 'interstellar particles' with a diameter of order 10 cm. If you wiggle your finger rapidly on the surface of the bath water between your 'tummy' and your knees you will make water waves with a wavelength of a couple of centimetres – short waves, analogous to blue light. Observe that these waves are readily reflected off your knees (scattering), and leave a clear zone just beyond them. Those waves which pass between your knees and rejoin near your feet are much diminished in height (absorption). Now rock your body back and forth (gently!) in the water. You will make long-wavelength (of order 30 cm) waves (analogous to red light). These waves pass almost without effect around your knees, very little diminished or scattered.

The analogy carries over from water waves to light waves. A light wave of wavelength 0.4 μm (blue light) which encounters a 1 μm interstellar particle is on the whole more likely to be absorbed or redirected than is a light wave of wavelength 0.7 μm (red light). If a spectrum of white light passes into an interstellar cloud, the blue light is selectively absorbed compared with the red light. The light which passes through the cloud is thus reddened (or actually de-blued!).

Atmospheric extinction

Reddening of light which passes through dust is a common terrestrial phenomenon. The Sun appears a light yellow when high above the horizon but at sunset appears orange or red, because of scattering of the sunlight by dust and by air molecules themselves. In dusty places – such as deserts in high winds, volcanoes during eruptions, or scrub during bushfires for example (Fig. 58) – the sunsets can be very red and very dramatic because of the large amounts of micrometre-sized dust in the atmosphere through which the Sun shines.

It is seldom that seaside sunsets are as beautiful since water droplets are usually much larger and do not act as selectively as smaller ones. Seaside sunsets generally appear orange, but in fine misty weather when the droplets of water suspended over the sea are small and act like small dust particles, seaside sunsets may be dramatically red.

The shift in colour of a star by the effect of the atmosphere as the star sets is called differential atmospheric extinction. It can be seen in Fig. 59, showing trails of setting stars. The star trails dim towards the horizon and become redder. The effect is easily measureable. In Fig. 60 is shown the increase in the colour index (see Chapter 1) of stars as they set towards the horizon (increased colour index means a redder star).

When astronomers measure the colour of a star by photoelectric photometry, the colour index which they measure is influenced by the particular value of the differential atmospheric extinction occurring when the star is observed. Astronomers use a correction curve, like those in Fig. 60, appropriate to the observatory from which the measurements were made. They adjust the measured colour index of a star to the value it would have if there were no atmosphere above the observatory. In this way the colours of stars as measured at different altitudes above the horizon and from different observatories may be compared meaningfully. It is always necessary to check the standard correction curve each night on which measurements are made in order to see if something unforeseen is upsetting it. In 1982 the standard curves were upset at all observatories around the world by dust ejected into the atmosphere by Mt Chichon, a Mexican volcano whose eruption far surpassed the more publicised earlier eruption of Mt St Helens in Washington. The dust was spread around the globe by upper atmospheric winds. It was a year before the atmospheric extinction curves returned to their standard form. Local effects may make it necessary to use different curves in the eastern and western halves of the sky. To overcome such uncertainties, astronomers seek the clarity of the air above high mountains.

Sunsets

Sunsets show the reddening of a bright star (the Sun) as its light traverses a large amount of atmosphere. In Fig. 61, we show not the Sun itself but its light

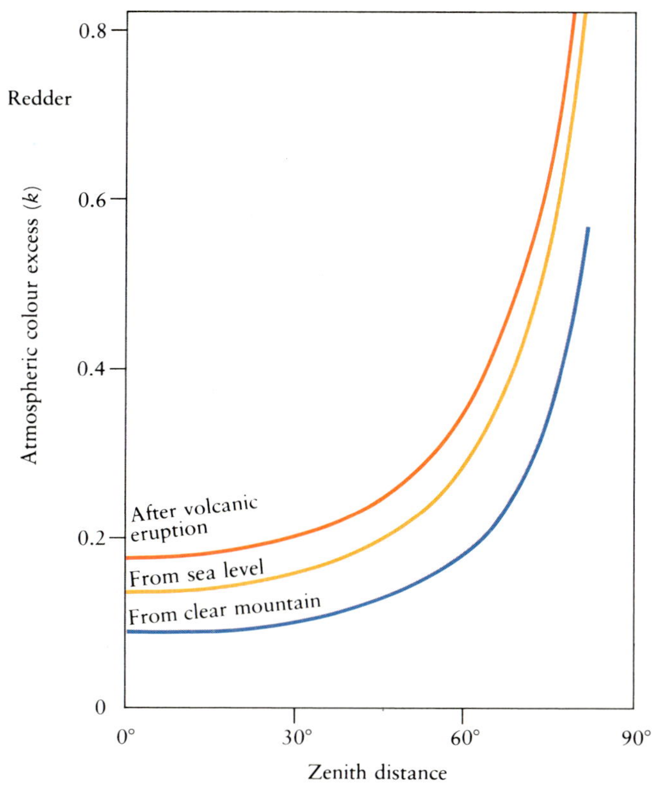

Fig. 60. *Extinction and zenith distance. The increase in colour index,* k, *is calculated for stars at increasing angles from the zenith. Horizontally viewed stars are at a zenith distance of 90°. Stars overhead are at zenith distance 0°, and even here where the line of sight passes through the shortest path of atmosphere, there is measurable extinction. Three curves are shown, one for a sea level site, one for a mountain site (both with clear sky, but the mountain top above much of the atmosphere), and one for a mountain site after the eruption of Mount Katmai in June 1912 which deposited large amounts of volcanic dust into the atmosphere (Abbot and Fowle, 1913).*

reflected off a large white object and the sky in the direction opposite to the Sun. In Fig. 61(*a*) (an hour before sunset) the observatory dome is white and the sky above deep blue. Half an hour later (Fig. 61(*b*)) the Sun has reddened and so the dome shows yellow. Fifteen minutes before sunset (Fig. 61(*c*)) the dome is orange. The sky above is still deep blue, but the horizon has become paler. At sunset (Fig. 61(*d*)) the Sun's intensity is so weakened by atmospheric absorption that the dome reflects a larger proportion of its light from the sky – it is not as red in the fourth picture as in the third. Beyond the dome, just above the horizon, can now be seen the *counter-twilight* of pink and yellow, the atmosphere in this direction illuminated by attenuated and yellowed sunlight like the dome. After sunset (Fig. 61(*e*)) the earth's shadow rises in the east, visible as a purple part of the atmosphere no longer illuminated directly by the Sun. By Fig. 61(*f*) (15 min after sunset) the purple has intensified and the dome has begun to redden again as it is illuminated by the twilight sky. An hour and a half after sunset (Fig. 61(*g*)) stars become generally visible and the violet light of the sky represents only the most readily scattered light. The last exposure (Fig. 61(*h*)), begun in the final minutes of astronomical twilight, shows the dome illuminated only by the dimmest reddest glow from the sky from where the Sun set.

Lunar eclipse

The same red colour on the dome in Fig. 61(*h*) can be seen on the Moon in Fig. 62. The photograph was obtained during a lunar eclipse, when the Earth lay between the Moon and the Sun. The parts of the Moon still illuminated by the Sun show white. If standing on the Moon just at the edge of the white area, an observer would see the Sun setting behind the Earth. At mid-eclipse, all the light reaching the Moon has passed through the Earth's atmosphere at almost grazing incidence with a pathlength twice that of any normal terrestrial sunset. So not only is the Moon when eclipsed extremely dark, it is (usually) very red as well. Its redness makes the eclipsed Moon a dramatic symbol of doom and it is forecast in the Bible by the prophet Joel (*Joel*, 2: 31) and the Apostles John (*Revelation*, 6: 12–13) and Peter (*Acts*, 2: 20) as a precursor to the Day of Judgement. Indeed, mediaeval and biblical annals commonly state the appearance of the Moon as blood red in order to indicate that it was in eclipse. The tradition that the Moon appeared like blood on the evening of Christ's crucifixion has been interpreted by C. Humphrey and G. Waddington of Oxford University as signifying that the Moon was eclipsed at that time.

This is consistent with the statements in all four Gospels that Christ died at nightfall on a Friday just before the Jewish feast of the Passover, which is always held at the time of a Full Moon when it is possible that the Moon may be eclipsed. (Note that an eclipse of the Sun is impossible at the time of Full Moon, so the darkness which the Gospels say fell at Christ's death was not due to a solar eclipse.) Humphrey and Waddington used their interpretation of the Moon's colour, in conjunction with the certainty that the crucifixion occurred during the decade that Pontius Pilate was procurator of Judaea, to search for lunar eclipses visible from Jerusalem on a Friday at Passover between AD 26 and 36. They fix the date of the crucifixion to that of the single lunar eclipse fulfilling these criteria, on Friday April 3, AD 33.

Steve Lee's time exposure (Fig. 62) of the eclipsed Moon captures its true colour, which is not an exaggerated 'blood red' but the colour of white sunlight minus all its blue and much of its green, in fact the deep orange colour of the sunset-illuminated dome in Fig. 61(*h*).

(a)

(b)

(c)

(d)

(e) (f) (g) (h)

Fig. 61. *The three hours of sunset. This series of pictures begins 1 h before sunset, with the final exposure (lasting 68 min) being terminated 2 h after the Sun had slipped below the western horizon, which was more or less behind the camera. The brilliant reds are due to multiple scattering of light from below the horizon and reach maximum colour saturation (but not brightness) long after the Sun has set.*

Fig. 62. *Lunar eclipse. Dramatic proof of the reddening effect of dust in our atmosphere is provided by this long-exposure photograph of the part-eclipsed Moon. All the light in the shadow area has passed through the maximum possible thickness of the Earth's atmosphere and is of low intensity. An exposure of several seconds was required to.record this picture. The effect of the Earth's atmosphere on astronomers on the Moon, anxious to learn more about the Sun's atmosphere during the total solar eclipse which they view at this time, must be distressing indeed!*

Chapter Four

Changing colours of the stars

Clusters of coloured stars

Discovered by the Abbé Nicolas-Louis de Lacaille in 1751 on his astronomical expedition to the Cape of Good Hope, the Jewel Box (NGC 4755) is an open cluster of stars near the Southern Cross, 7800 l.y. from the Sun (Fig. 63). Its central star is Kappa Crucis. (The most prominent stars in a constellation are traditionally identified by Greek letters, the brightest being Alpha (α), the next brightest Beta (β) and so on.) Kappa, the tenth brightest star in Crux, is easily seen to be reddish in a modest telescope; the remainder of the stars in the cluster give various impressions of colour, all more or less white or blue. NGC 6520 (associated with Barnard 86, Fig. 57) is another Jewel Box-like cluster with blue stars in association with yellow ones. NGC 3293 (Fig. 64) is another cluster which shows the same distinction in colour between the brighter red stars and the moderately bright blue stars. The famous open cluster of the northern winter sky, the Hyades, is yet another.

The first astronomer to remark on the curious fact that the stars in clusters come only in certain combinations of colour and brightness was Ejnar Hertzsprung. He was led to this conclusion in 1911 by comparing the Hyades with the stars of the Pleiades cluster.

The individual stars of the Pleiades, so conveniently near to the Hyades that the two clusters beg to be compared, have been well known from antiquity. Most people can see the six brightest with their unaided eyes. In classical literature the stars have always been associated with seven sisters or seven doves. The keen-sighted can see more than seven stars however, and in 1579 Kepler's teacher Mästlin drew a chart showing 11 of the Pleiades stars in what we now

Fig. 63. *The Jewel Box. This cluster of stars was named the Jewel Box following its description by Sir John Herschel: 'The stars which compose it, seen in a telescope of diameter large enough to enable the colours to be distinguished, have the effect of a casket of variously coloured precious stones'.*

Fig. 64. *NGC 3293. Visible here is the same distinction of colour between a bright red star and the surrounding whiter stars as can be seen in the Jewel Box.*

Fig. 65. *Pleiades. Produced from W. Miller's original Palomar exposures, this photograph shows the blue stars of a young star cluster embedded in blue nebulosity. The nebulosity is streaked, apparently by interaction with the interstellar magnetic field.*

know to be their correct places: this was 30 years before the invention of the telescope with which Galileo could check Mästlin's feat. Galileo in fact counted 36 member stars. Presently it is estimated that there are 300 to 500 members, with the most complete catalogue, published in 1958, listing the brighter 262. Hertzsprung measured the brightness of many of these stars on two photographs of the Pleiades obtained through blue and green pieces of glass, and, forming the ratio of the blue and green brightnesses for each star, he was able to construct its colour index. He confirmed some of his measurements by looking at the spectra of the brightest stars and measuring the ratio of green and blue brightnesses directly. He was then able to correlate the colour indices of the Pleiades stars with their brightnesses. This led him to his fundamental discovery.

The Main Sequence

The Pleiades do not include any *bright* red or yellow stars – all the bright ones are conspicuously blue on colour photos (Fig. 65). There is in fact a progression from the bright blue stars in the Pleiades to *fainter* redder ones. This correlation of colour and brightness simultaneously is called the Main Sequence, and Hertzprung's own diagram of the Main Sequence of the Pleiades is shown in Fig. 66.

Hertzsprung was eased in his discovery by the fact that the stars in the cluster were at a common distance so that the dimming effects of distance (and interstellar absorption) were equal for each star. A cluster star which appears four times as bright as another is indeed four times brighter and not at half the distance. It is much more difficult to examine the correlation between colour and luminosity for stars in general, since they are at varying distances from the Earth. Only if the stars' distances are known can the Main Sequence be found. In fact the Main Sequence was indeed discovered in this way virtually simultaneously with Hertzsprung's diagram by Henry Norris Russell. Russell studied those nearby stars whose distances had been determined by a celestial surveying technique called the method of parallaxes. This involves repeatedly sighting on a nearby star from the two ends of the largest baseline available to us, namely the diameter of the Earth's orbit around the Sun (Fig. 67). The position of the star against the background of distant galaxies shifts in much the same way as a finger's position shifts against the background when viewed by each eye winking open in turn. The shift of distant stars is small because sightings from the extremeties of the baseline are parallel. For the nearest stars the shift can be up to about 1 arc second. The *angle of parallax* is the angle subtended at the star by the radius of the Earth's orbit, and a star which has a parallax of 1 arc second is said to have a distance of 1 *parsec* (pc). An angle of 1 arc second represents 1/1800 the apparent diameter of the Moon, or the apparent diameter of a 1-cm diameter coin at a distance of about 2 km. The shift of a star with a parallax of 1 arc second is the same as the shift in position of a needle at 15 km when viewed with each eye alternately. There are in fact no stars as near as 1 pc

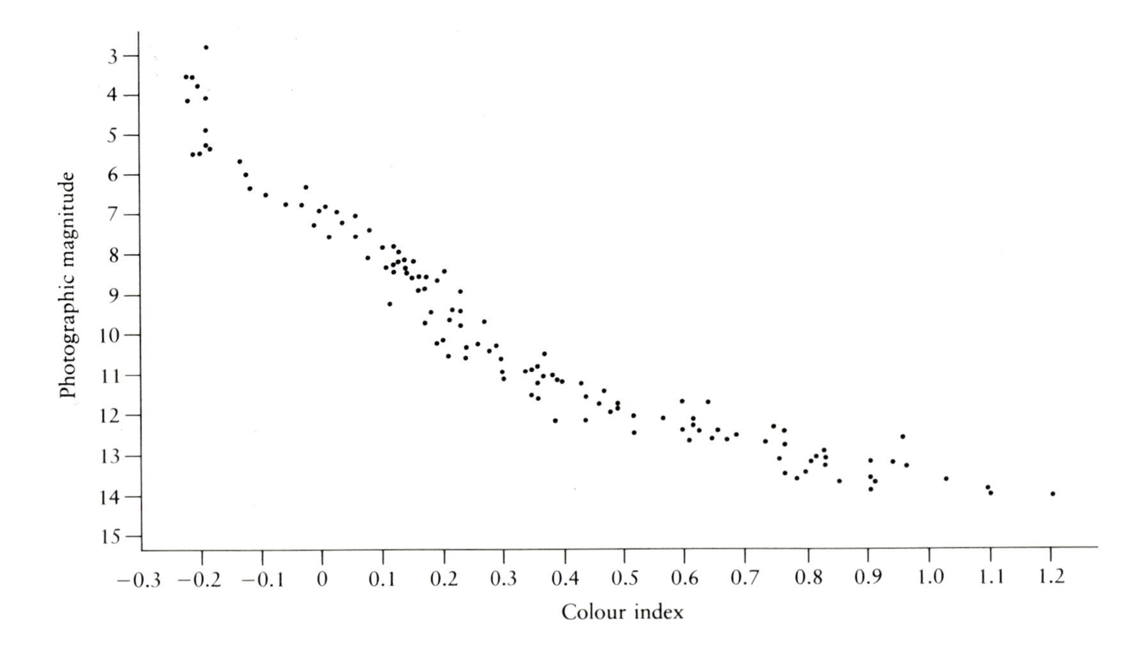

Fig. 66. *Pleiades' Main Sequence. Hertzsprung's own colour–magnitude diagram of the Pleiades shows by measurement what is apparent in Miller's photograph. There are bright blue and white stars in the Pleiades (upper left) and plenty of fainter yellow and red ones (lower right), but there are no bright yellow or bright red stars: the upper right-hand corner of the diagram is empty.*

and the angle of parallax is more typically 0.01 arc second (the apparent diameter of the same coin at 200 km distance). A star with this parallax would be at a distance of 100 pc.

In 1913 Russell could use the measurements of the parallax of just over 200 stars to calculate their distances and to determine the brightness which they would all have if they formed a cluster at a standard distance from us of 10 pc (this figure, equal to 32.6 l.y., is an arbitrary, conventional choice). The magnitude of a star corrected to 10 pc is called its absolute magnitude. He then plotted the absolute magnitude of his sample of stars against their spectral type (Fig. 68). As we saw in Chapter 1, spectral type is correlated with colour index. Thus Russell was plotting essentially the same diagram that Hertzsprung was, but for the stars which by chance were found in the neighbourhood of the Sun. There is no real difference in the diagrams and the name of Hertzsprung–Russell diagram (H–R diagram) commemorates the two astronomers equally.

The giant branch

As well as the sequence of stars which runs from bright, blue ones to faint, red ones (the *Main Sequence*, on which most stars lie), there was also in Russell's diagram another sequence of stars which ran from bright blue to bright red. Such stars were relatively rare, with just a few or even none in a cluster of several hundred stars. This part of the H–R diagram became known as the giant branch, jutting out from the Main Sequence. The stars on it were called giants, because they were bigger than Main Sequence stars of the same colour. The red giants, for instance, were ten magnitudes brighter than the red stars on the Main Sequence. Ten magnitudes brighter means 10 000 times brighter. Since all red stars have more or less the same temperature, the red giant stars must have 100 times the diameter of the red Main Sequence stars. This justifies their name of *giant* stars and the contrast of sizes suggested the name of *dwarf* stars for all those lying on the Main Sequence.

The meaning of the H–R diagram

It was natural to think of the path of stars in an H–R diagram as a progression through which stars passed in their lifetime. This is exactly how Russell interpreted his discovery. He thought of a star as a gaseous body which contracted as it grew older. Indeed he believed that the radiated energy of a star came from its gradual shrinking under gravity and that the general trend of change in a star must be from large sizes to smaller. He recognised that there might be some stages in a star's life during which energy might be released by some other phenomenon – nuclear energy – but he thought that this would be a passing stage in the career of a star and that it would still be true in the long run that stars got smaller. The stars in the H–R diagram are distributed in size from largest to smallest from the red giants, along the giant branch to the blue stars, and then down the Main Sequence to the red dwarfs, so he thought of the H–R diagram as the pathway of stars which have been caught at various stages in their life. The life of a star was one in which it contracted overall, first heating from red hot to white hot as a giant and then cooling to red hot again as a dwarf.

Russell had done the straightforward thing in interpreting the H–R diagram. If we were to photograph London from the air during the rush hour we would notice a main sequence of pedestrian commuters in line across Waterloo Bridge and would rightly conclude that they were in motion along the sequence. The commuters' main sequence over the bridge, we could notice, is fed by branches emanating from Waterloo Station. We could confirm the pedestrians' motion by taking a second photograph a few minutes later. However, an astronomer's lifetime study of a star can only constitute a snapshot of a stage in its development. In the H–R diagram we have but a single snapshot of stars with generally no chance to take another snapshot to see the changes.

On closer examination of our photographs, however, we would notice a phenomenon which produced a different kind of sequence. At the railway station there would be rows of commuters lined up patiently on the platforms waiting for a train. Eventually a train would arrive and they would pass into its numerous doors and be rapidly taken away. Their place would be taken by later arrivals who joined the sequence of commuters on the platform from the numerous entrances to the station.

The commuters on the platform thus form a sequence but do not progress along it. The platform is the place where they spend a long time waiting; it is here that we are most likely to find them. In fact the

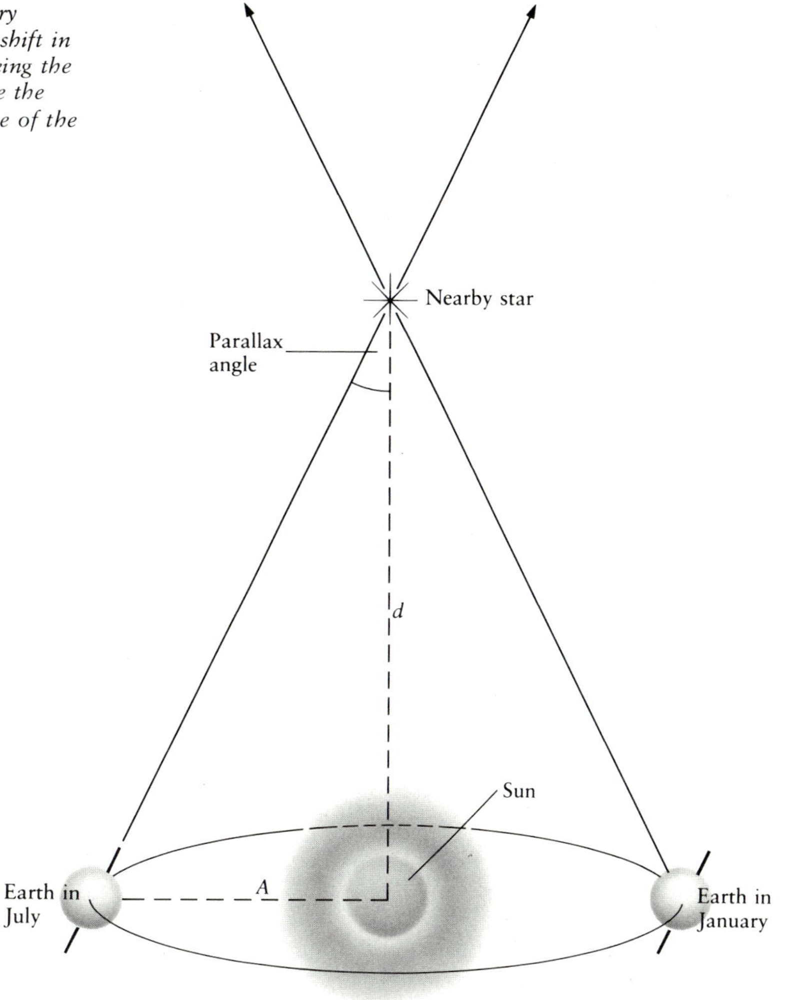

Fig. 67. *Parallax. A nearby star is viewed in January and July. Earth-bound astronomers can see the star shift in position relative to background galaxies, the shift being the parallax. By trigonometry, astronomers can calculate the distance,* d, *of the star because they know* A, *the size of the astronomical unit.*

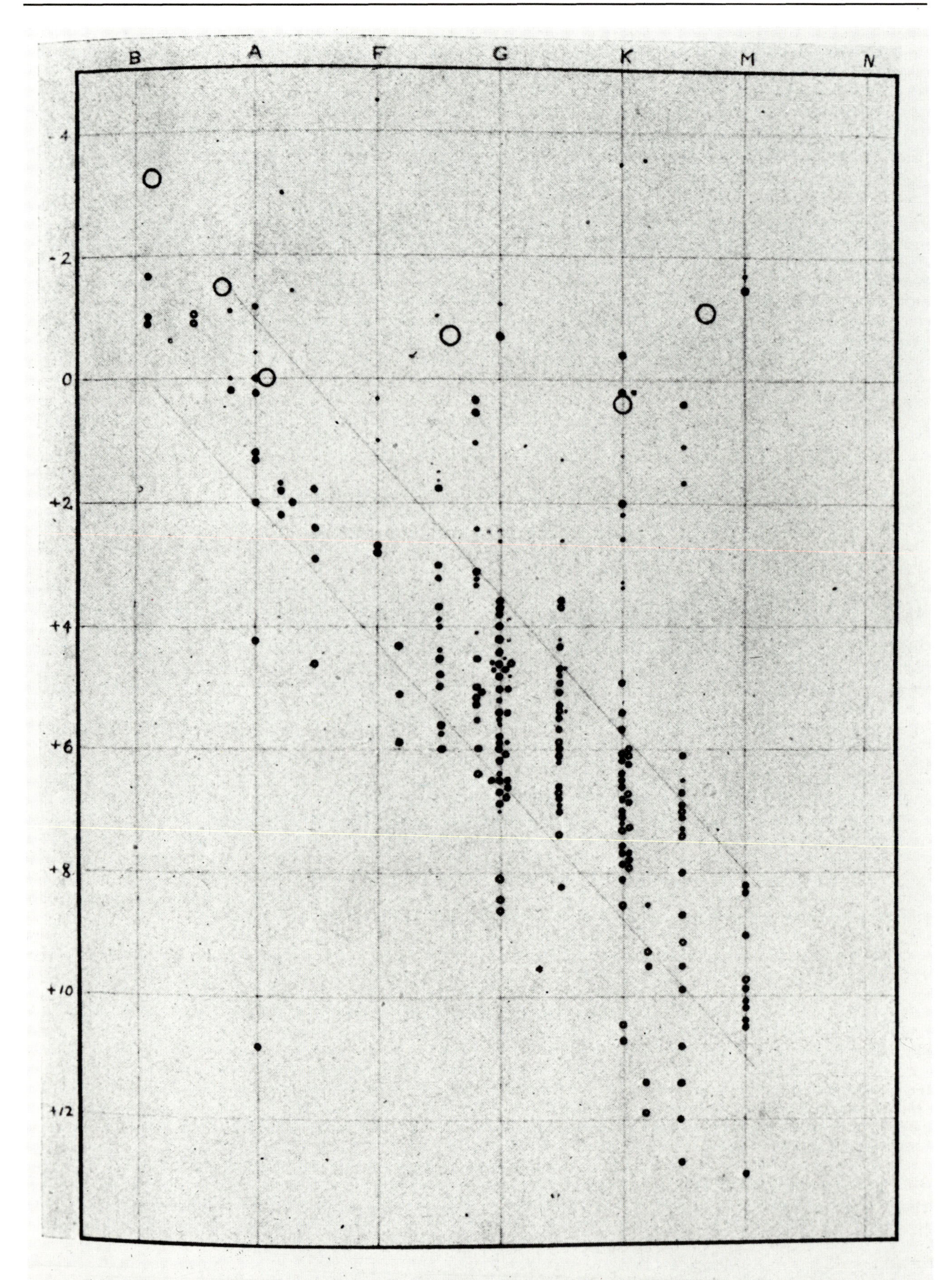

FIGURE 1.

Main Sequence of stars in the H–R diagram is a holding area of this kind, and not, as Russell thought, a pathway. It represents the status of all stars during the majority of their active lifetime while they are shining.

All stars are in a balance between the force of their own gravity and an inner pressure. Gravity tends (as Russell knew) to make the stars contract, and the outward surge of energy which they create by nuclear reactions in their interior creates the pressure. The balance between these inward and outward forces restricts the normal structure of stars to the narrow combination of surface temperature and brightness which is called the Main Sequence. A star of a given mass spends most of its life near one position on the Main Sequence. The least massive stars are the least luminous and the coolest, and so are red dwarfs; the most massive stars are bright and hot and are blue dwarfs.

The balance in a star

Since a star is in balance between its internal pressure and gravity, the star remains a dwarf for only as long as the balance is maintained. The force of gravity lasts indefinitely but the pressure does not. The pressure is the consequence of energy generated deep inside the star. At its centre, at temperatures of millions of degrees, the massive nuclei of atoms coalesce and synthesise even heavier nuclei. For instance, four hydrogen nuclei combine to make a helium nucleus in a fusion reaction. This releases energy in the form of heat which travels outward from the star's centre. Because the energy jostles its way out of the star, travelling almost as much inwards as outwards, it makes slow outward progress. In fact the energy diffuses out of the star over tens of thousands or even millions of years. It is the jostling of the material of which the star is made by the passage of the energy outwards which supports the body of the star against the force of its own gravity and stops it collapsing to a point.

The travel of the star's energy outwards shows that the temperature is highest at the centre of the star and lowest towards the outside: one does not expect heat to pass from a cool body to a hotter one, only from a hot body to a cooler one, so the outside of the star must be cooler than the inside. The outer limit of a star is rather hard to define – for instance, our Sun's atmosphere extends far past the Earth. (The Earth orbits within the Sun's gases, as proved by the existence of the aurora which is caused by the collision of parts of the Sun's atmosphere with our atmosphere.) What's more, the temperature of the material of the star varies, let us say from 10 million degrees in the middle of its nuclear reactor to the inert cold of interstellar space, which begins somewhere at the edge of the solar system (see Fig. 69). How then can we talk of '*the* temperature' of a star?

The answer lies in the 'jostling' process by which energy moves out of the star. In the middle of the star a given packet of energy travels for a tiny distance before hitting material in the star and being deflected, maybe back to the centre again, but generally outward. Towards the outside of the star, the material is thinner. As the packet of energy travels outwards, its chances of travelling longer distances increase. There is a layer at which the packet of energy has quite a good chance, say an even chance, of escaping without any more interaction with the material of the star – the star has become more or less transparent. This is the layer we call '*thê* surface' of the star, because it is the part we see – the heat or radiation which comes from here carries with it the distinguishing characteristics of this layer, simply because nothing further happens to the radiation.

What we see are the tops of huge circulating convection currents, rather like the mud of some volcanic pool, constantly rising, radiating and falling again to be re-energised and to reappear. The high-resolution picture of the Sun's surface (Fig. 70) shows the cool edges of these cells, each several hundred kilometres across. These cells are literally the surface of the star we see. It is at this layer that the colour of the Sun arises, 700 000 km above its active centre, carrying the history of thousands of years ago, when the energy was released at the star's interior.

Fig. 68. *Main Sequence of nearby stars. Russell plotted the absolute brightness* versus *the temperature of nearby stars and found the Main Sequence of the stars independently from Hertzsprung. Modestly he at first declined any honour for this discovery, asking what credit there was, when only two quantities about stars were known, in plotting one against the other.*

Changing colour and brightness

The conversion of energy from hydrogen to helium cannot last forever – the hydrogen in a star will run out. In the Sun for instance 5 million tonnes of hydrogen are converted into energy *every second.* Even the astonishing mass of the Sun (2000 million million million million tonnes) will be depleted after literally thousands of millions of years. One might imagine that the blue dwarfs, being more massive stars which have more fuel available, would last longer, but the reverse is true because the massive stars burn hotter and brighter and release their energy more quickly. Their lifetime on the Main Sequence is shorter than the Sun's, just as a profligate millionaire may go bankrupt quicker than a poorer but less spendthrift man.

When the hydrogen is finally exhausted the internal make-up of the star changes and it seeks a

Fig. 69. *(left)* *Solar corona. The corona of the Sun becomes visible at a solar eclipse, as at this one on 31 July 1981 visible from Kazakstan in the USSR. The solar atmosphere can be traced on this picture to a distance of a couple of solar diameters, but extends well past the Earth to the outer reaches of the solar system.*

Fig. 70. *(below)* *Solar granulation. The tops of convection cells are visible as granules on the surface of the Sun in high-resolution pictures obtained in the clearest mountain air. The granulation surrounds a group of sunspots in this picture from the Swedish solar telescope on La Palma.*

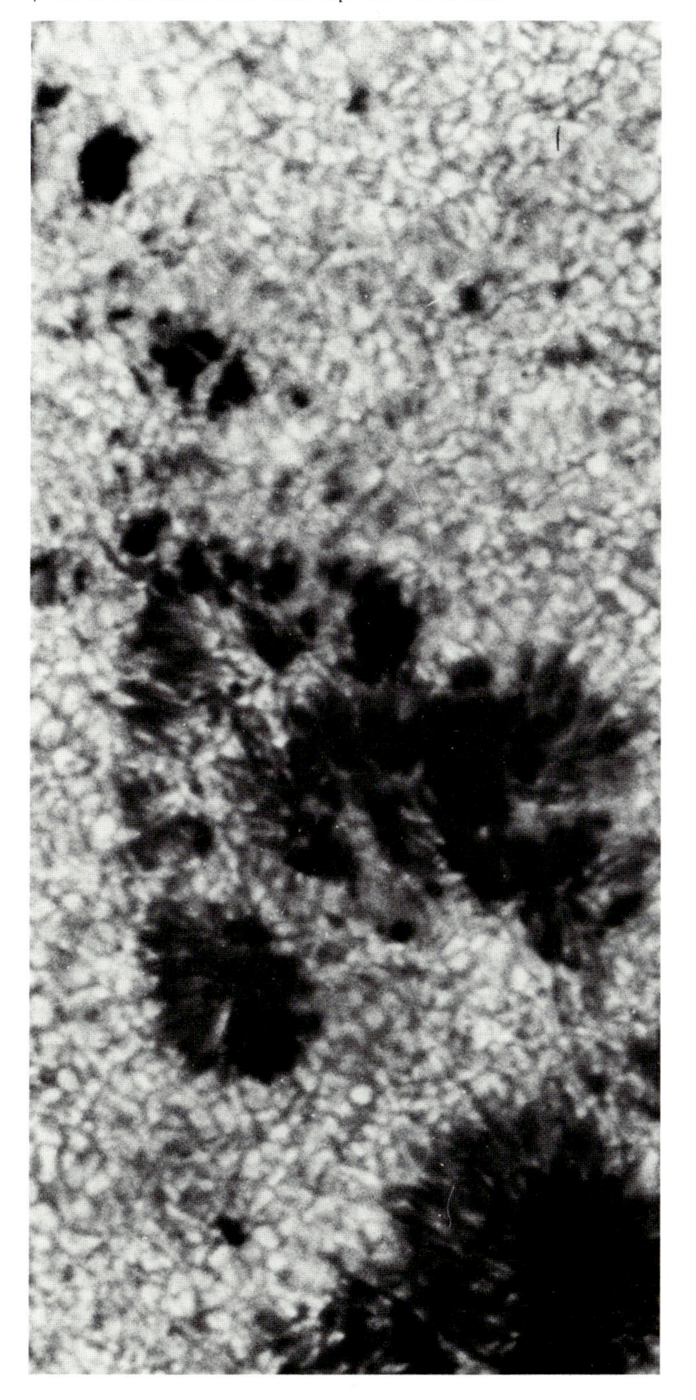

new balance between energy generation and the force of gravity. These changes displace a star from the Main Sequence and its properties may quickly change as it evolves into a red giant. Because the blue dwarfs move into the giant region first, the Main Sequence of a cluster in which the brightest stars have exhausted their nuclear fuel curves over at the bright tip and a few red giants appear.

Now we know why the Jewel Box, Hyades, and the other clusters whose photos appear on pp. 79 and 80 contain bright red giants, and why the Pleiades contain none. The Pleiades are younger than the stars of the Hyades and the Jewel Box. The Pleiades are 50 million years old. The Hyades are, however, more than 1000 million years old.

As time passes, the upper part of the Main Sequence of a cluster of stars is eaten away as the stars become red giants. The red giant stage does not last long and the star rapidly evolves into a very faint star called a white dwarf. Such stars are often very hot but are always very faint indeed, because they are so small. Effectively they disappear from view. Old clusters which may be tens of billions of years old (the so-called globular clusters) have no blue dwarfs. They even have no yellow dwarfs, because these, too, have evolved off the Main Sequence. The stars which have most recently left the Main Sequence have become red giants, so that predominantly all the brighter stars in the cluster are red (Figs 63, 64). A few stars have sped from the red giant stage and are on their way to become white dwarfs. They evolve through an area of the H–R diagram called the horizontal branch, and are briefly blue or yellow giants. They show on the photo as whiter stars.

Disturbing the balance

The balance of the structure in a star is delicate but self-correcting; it usually adjusts to the slow exhaustion of the star's nuclear fuel in a quietly progressive way. The star's path in the H–R diagram as it evolves off the Main Sequence to become a red giant is, it is true, an accelerating one, but it usually takes place in a continuously changing manner as the star grows in size like an inflating balloon.

However, just as the progressive increase in size of a balloon may be violently disturbed as it inflates to meet a pin, so there are circumstances in which the continuous growth of a star may be suddenly altered. One such circumstance occurs when an evolving star is a member of a double star system. Then, as the brighter star grows in size and becomes a red giant it may overflow onto the adjacent star. The star may pass on to become a white dwarf more quickly than usual, while the other star grows in mass because it has accreted the overflow from the red giant. D. Lauterborn (1970, 1971) has calculated the evolution of a double star system intended to imitate the double star Sirius. His model consists originally of two stars A and B of masses twice that of the Sun and five times that of the Sun and having a separation 300 times the solar radius. The larger star becomes a white dwarf (like Sirius B) circling a blue dwarf star (like Sirius A) whose mass has increased to 6 solar masses.

Lauterborn made this calculation in order to investigate a remarkable claim, based on the study of historical records, that the colour of Sirius has changed from the red of Sirius B in classical times to the present colour of Sirius A, which all agree is white (see Table 6). (The two stars are close together and cannot be distinguished as individuals by the unaided eye.)

The colour of Sirius*

The first man to draw attention to the possibility of a change in the colour of Sirius was Thomas Barker, a prolific but little-known eighteenth-century scientist, writer and theologian who, a vegetarian, lived to the age of 88. Barker (1760) pointed out that the colours of the stars and planets were generally the same as the ancients had observed. Ptolemy, who compiled (AD 140) the catalogue of stars known as *Almagest*, marked as distinctive the colour of only six of them: he called them *hypokirros*. Although the word itself means *somewhat yellow*, five of the stars are so reddish coloured that Liddell & Scott's classical Greek dictionary notes the word as meaning *red* when applied to stars. The six stars are listed in Table 8. We can use the Minnaert colour scale (Table 6) to interpret the $B-V$ colour index of the stars, to say that five are what Minnaert would observe as pure yellow to yellowish red. The outstandingly odd member of the sextet is Sirius, the brightest star in Canis Major and known as the Dog Star. Because it is the brightest star in the sky its colour should be easily seen. Now, even to an uncritical eye, it is plainly white. Because Ptolemy called it *hypokirros*, the implication is that the colour of Sirius has changed in 2000 years from reddish to white. Barker added other evidence by giving quotations from classical authors which tended to imply that Sirius had a red colour. The clearest seem to be the following by Seneca (4 B.C.–A.D. 65): 'The redness of the Dog Star is more burning; that of Mars is milder; Jupiter is not red.' and by Horace (65–68 B.C.): 'The red Dog Star divides its children.'

The comparison with Mars and Jupiter seems straightforward enough and implies that Sirius then had a colour index ($B-V$) greater than 1.4, describable as orange, or redder. Horace's statement,

* We have been helped in the arguments in this section by correspondence with Robert Temple, author of *The Sirius mystery*, after his extensive historical research on the subject.

TABLE 8. STARS WHICH PTOLEMY CALLED *HYPOKIRROS*

Name	Star	$B-V$	Minnaert colour
Arcturus	α Boo	1.24	Pure yellow
Aldebaran	α Tau	1.53	Orange*
Pollux	β Gem	1.00	Deep yellow*
Antares	α Sco	1.83	Orange
Betelgeuse	α Ori	1.86	Yellowish red*
Sirius	α CMa	−0.01	White

*Not observed by Minnaert; colour inferred from ($B-V$)

which is about the position of Sirius relative to fainter nearby stars, is unambiguous about its colour.

Many astronomers accepted Barker's arguments. Sir John Herschel (1839) had an explanation for Sirius' red hue. Responding to an enquiry about the changes in brightness of Eta Carinae, he hypothesised that space contains 'opake matter' in clouds and that the clouds in travelling in front of stars cause them to change their brightness in a haphazard manner. He went on to relate this theory of variable stars to Sirius' colour: 'Of the nature of these super-atmospheric clouds of course no conjecture can yet be formed, but some argument for their being of a material nature *may* be drawn from the strange observation of Ptolemy that Sirius was in his time one of the 6 red stars . . .' says Herschel, emphasising his overall caution by italics. 'It seems much more likely that a red colour should be the effect of a medium interposed, than that in the short space of 2000 years so vast a body should have actually undergone such a material change in his physical constitution.'

Schiaparelli (1896) and Simon Newcomb (1902) did not believe in the colour change of Sirius from red to white, because they endorsed a version of stellar evolution in which white stars cool to red ones ('as a piece of iron cooling from white heat'), the reverse change. Newcomb calculated that the time scale for the cooling would be tens of millions of years not thousands, and could not believe the rapidity of the change. It is interesting that the astronomer B. Paczinski (1970), after listening to Lauterborn's exposition of his theory of the change at a conference at Elsinore in Denmark, objected with a similar argument to Newcomb's, namely that even if Lauterborn's scheme produced a white dwarf star quickly, it takes many millions of years for a newly born white dwarf star to cool to become as Sirius B looks now, with the rate of cooling slowing to a billion-year time scale by then.

These arguments, based on 'known' theories, are of course dangerous, since theory may change rapidly. Schiaparelli's and Newcomb's arguments about the direction of the change of colour would not have been valid a few years afterwards when Russell's theory admitted the possibility of heating from red to white as a star contracted along the giant branch of the H–R diagram (p. 84) before cooling down the Main Sequence; and it has to be pointed out that the initial cool-down time for white dwarfs was thought to be 100 million years in 1969 but was later amended to 10 million years, following the discovery that cooling of white dwarfs takes place primarily by loss of neutrinos, not photons. D'Antona and Mazzitelli (1978) argued that no-one can exclude the possibility that the theories are wrong: perhaps Sirius B has a hydrogen shell which could be unstable and appear like a red giant for a few thousand years before collapsing back to look like an ordinary white dwarf again.

It is best to look back at the original evidence for the change in Sirius' colour, and many astronomers have done so: Barker's claim was argued with great enthusiasm in the latter half of the nineteenth and early twentieth centuries when astronomers knew much more Greek and Latin than now. Barker's notion was first popularised by Baron Alexander von Humboldt (1845), in his mammoth work *Cosmos*, who left no doubt as to his view: 'Sirius is the one example of a star historically proved to have changed colour'. It was probably from the English edition of this that T.W. Webb (1851) culled his remarks on the colour of Sirius in *Celestial objects for common telescopes*. 'Its colour has probably changed. Seneca called it redder than Mars; Ptolemy classed it with the ruddy Antares. I see it of an intense white, with a sapphire tinge, an occasional, probably atmospheric, flash of red.' Both popularisers accepted the idea that Sirius had been red, although by the fifth edition (1893) of Webb's highly successful book, the editor R.E. Espin had inserted a caution into the text: 'Lynn, however, doubts the construction put on the evidence'. Lynn (1887) attributed the classification of Sirius as *hypokirros* to a transcription error.

Enter Dr See

Barker's arguments were taken up by one of astronomy's most colourful figures, the notorious Dr Thomas Jefferson Jackson See. T.J.J. See was interested in geodesy and in celestial mechanics. In his early career he observed double stars and planets and seems to have been an able user of the telescope: he may even have glimpsed the craters on Mercury, the existence of which were not proved until the Mariner 10 flyby of the planet in 1974 (Baum, 1979, Lankford, 1980; Hoyt, 1981). However, 'the trajectory of his career was that of steady decline', according to John Lankford. In 9 years, he moved rapidly on from a PhD at Berlin to become an instructor at the University of Chicago, a member of the Staff at Lowell Observatory, and an astronomer at the US Naval Observatory, first in Washington D.C. then Annapolis and ending up in the Navy Yard, Mare Island, California; here he remained until he retired. He sought fame in the newspapers, in later years arguing publicly against the theory of relativity. He achieved notoriety among his peers who recognised him as an egotistical charlatan. T.J.J. See was satirised by H.H. Turner (1899), a regular contributor to *The Observatory*, a monthly review of astronomy which is still to this day noted for the beady eye and sharp tongue with which it notes and lashes those who fall short of its own exacting standards. It announced in 1899 a new *fundamental law of increase of gaseous reputation* (parodying the title of See's (1898) paper on 'The fundamental law of temperature for gaseous celestial bodies'):

> *Let T be the scientific reputation that may be established by any means whatever, on the condition that it may be of a perfectly gaseous nature. Let J be the value which the investigator places upon popular newspaper or magazine notoriety, and J_1 that which he attaches to sound reputation as gauged by the good opinion of men whose good opinion is worth having. Further, let C be the amount of his egotism, not to be regarded as constant but as rapidly increasing at an enormous ratio, approaching asymptotically to the limit of mental alienation. Then we have*
>
> $$T = \frac{J}{J_1} C$$
>
> *It is manifest that under the conditions named this function may assume an infinite value, as illimitable as the boundless sea, since it is composed of two factors, one a ratio which may become as large as we please by increasing the numerator and diminishing the denominator, each without limit, and the other factor, C, is a quantity that by hypothesis expands indefinitely as a necessary consequence of the operation of the law.*

With his customary excess, See (1892, 1927) took up Barker's argument that Sirius had been red. In the second of the two papers (the second repeats much of the first of 35 years earlier) he quotes 20 classical authors (Table 9) whom, he says, support the argument, although only a handful could be said clearly to refer to the red colour of Sirius. The nadir of the argument is a reference to the incredibly obscure author Sextus Pompeius Festus who says that at the *Floralia* red-haired dogs were sacrificed to placate the Dog Star (Table 9, No. 16); 'and one naturally asks why ruddy dogs rather than dogs of any other color?'

TABLE 9. T.J.J. SEE'S EVIDENCE ON THE COLOUR OF SIRIUS

Author	Sirius
1. Ptolemy	*ὑπόχιρρος*
2. Geminus	*πύρινος* (multitude *αἰθέρια*)
3. Seneca	*Acrior sit Caniculae rubor, Martis remissior, Jovis nullus*
4. Pliny	*Ardore Sideris, Sirio ardente*
5. Aratus	*ποιχίλος*
6. Cicero	*Rutilo cum lumine claret fervidus ille Canis.* – 'with ruddy light fervidly glows that dog'
7. Germanicus	(a) *Canis ore timendo*
	(b) *ore vomit flammam,*
	(c) *urgetur cursu rutili Canis ille per aethram*
8. Avienus	*Multus rubor imbuit ora*
9. Theon	*ποιχίλος* – 'highly coloured'
10. Columella	*Sirius ardor*
11. Horace	*Rubra Canicula*
12. Virgil	*Ardebat in coelo*
13. Manillius	*Rabit suo igne*
14. Euripides	*πυρὸς φλογέας*
15. Ap. Rhodius	(a) *οὐρανόθεν Μινωίδας ἔφλεγε νήσους Σείριος*
	(b) *ἱερῆες ἀνατολέων προπάροιθε χυνὸς ῥέζουσι ψυηλάς*
16. Festus	(a) *Rutilae Canes immolabantur ad placandum Caniculae Sidus*
	(b) *Rutilae Canes, id est non procul a rubro colore*
17. Ovid	*Pro cane Sidero canis hic imponitur arae*
18. Ateius Capito	*Rutilae Canes immolantur causa Sideris Caniculae*
19. Hesiod	*Σείριος ἄζει*
20. Homer	(a) *φέρει πολλὸν πυρετόν,…*
	(b) *ἀχάματον πῦρ, ἀστέρ' ὀπωρινῷ ἐναλίγχιον*
	(c) *οὔλιος ἀστὴρ* – 'deadly star'

Was Sirius really seen to be red?

Most of the authors cited by See in Table 9 refer to the burning or fiery nature of Sirius; these references are ambiguous because they may simply have in mind the brightness of Sirius. It is, after all, the brightest star and in clear skies over desert mountains shines with an intensity almost painful to look at. Those classical authors who clearly refer to the redness of Sirius include Ptolemy, Seneca and Horace (as already cited by Barker). Cicero is another, but he is merely translating sentences by Aratus (315–240 B.C.), who describes Sirius with the word *poikilos*. See says this means red or ruddy as Cicero, apparently carelessly, implies; but what does the original word mean? *Poikilos* is listed in the dictionary as meaning 'variegated, mottled, pied, dappled, wrought in various colours, changeful, various, diversified, manifold'. Aratus was a poet whose *Phaenomena*, from which the lines are taken, is a long astronomical poem about the stars, the constellations and meteorology. He shows evidence of having actually studied the sky and from his description of Sirius appears to have noticed the same scintillation or twinkling of stars as his modern counterpart, Robert Browning (*My star*):

All that I know
Of a certain star
Is, it can throw
(Like the angled spar)
Now a dart of red,
Now a dart of blue;
Till my friends have said
They would fain see, too,
My star that dartles the red, and the blue!

Browning's verse correctly implies the mechanism for the twinkling colours of stars. Light from them passes into our atmosphere of air and is refracted. The angles by which the various coloured rays are refracted differ for each colour and spread the star image into a rainbow of blue to red. Normally this cannot be seen with the unaided eye, but it is possible for a bubble of hotter-than-average air to catch the red or blue rays and deflect them more than normal to direct a dart of red or blue into the eye. This scintillation is what Aratus was probably describing with the word *poikilos*, rather than the redness of Sirius.

It may well have been that Seneca also had in mind the same phenomenon. In context, Seneca's passage of the *Quaestions naturales* reads (John Clarke's 1910 translation):

> *And little wonder if the earth's evaporation is of all varied kinds. Why, even in the heavens the colour of objects does not show uniform; the red of the Dog-star is brighter, that of Mars duller; Jupiter has no red, his sheen is prolonged into pure light.*

The reference to the colours of the stars is an aside in a passage on vapours rising from the Earth. There is a mix-up between colour, intensity and scintillation, with the steady light of Jupiter contrasted with the variable colour of other objects – because of, or like the Earth's vapours? The best that could be said of the total passage is that it is confusing but appears to refer to scintillation.

Horace is another poet on See's list who refers to Sirius as red – *rubra Canicula* can mean nothing but *the red Dog star*. Even See (1927) admits that Horace, a poet notorious for not knowing anything about natural phenomena, was scarcely an astronomical authority. The same kind of remarks are made by See himself about Hyginus who, in a passage somewhat muddled about the names of the stars, implies that Sirius is white. See left Hyginus out of his table of evidence, because he did not fit, but put Horace in, because he did.

We are left with Ptolemy's apparent observations of the sextet of red stars. In *The crimes of Claudius Ptolemy*, Robert Newton has described how Ptolemy fudged observations of the positions of planets etc., calculating them from theory. It is hard to see why Ptolemy should falsify the colour of Sirius, but this does cast doubt on the 'observation'. Another possibility proposed by Schjellerup (1874), Lynn (1887), Schiaparelli (1896) and Temple (1975) is that *hypokirros* is a mistaken transcription of the name of Sirius itself. Stars in *Almagest* are described by their positions in their constellations as well as their names. Thus: Alpha Boo is *the fiery star called Arcturus, between the thighs*; Alpha Leo is *the one at the heart, called Regulus*; and Alpha CMi is *the bright star in the hind parts, called Procyon.* Does the entry for Alpha CMa say *the brightest and reddish star, called the Dog, in the face*? or does it say *the brightest star in the face, called Sirius and the Dog*? The latter makes more sense. The Greek words *Sirius* and *hypokirros* are not alike and See (1927) claimed that it was out of the question that there could have been a transcription error. It is equally unlikely that Ptolemy would have referred to the brightest and most famous star without mentioning its name, the name of other stars being given throughout.

Finally, let us note that Chinese observations by Sima Qian in the first century B.C. call Sirius white (Fang Li-Zhi, 1981), so that the classical observations are unconfirmed in other cultures.

Many astronomers have puzzled over the change of colour of Sirius from red to white in 2000 years (Herschel, 1839; Newcomb, 1902; Eddington, 1911; Osthoff, 1927; Gundel, 1927; Kopal, 1959; Lauterborn, 1970, 1971; Rakos, 1974; Lindenblad, 1975; Maran, 1975; D'Antona & Mazzitelli, 1978). They may have been stimulated to their work by a slip of a pen, a careless translation, an unconsidered adjective, and a charlatan's egotistical overenthusiasm. The change in colour of Sirius seems to have been a red herring to astronomy, however real the long-term changes of the colours of stars.

Dumbell Nebula. A planetary nebula which has a double-lobed structure, the Dumbell is centred on its progenitor star.

Chapter Five

Nebulae from stars

Mass loss in stars

Like some human beings, stars seem particularly sensitive about their 'weights' at certain phases of their lives. As with us, 'weight consciousness' and the shedding of mass is often associated with growing older. However, stars also adjust their internal balance at the beginning of their lives. When a star loses mass it can do so in a variety of ways, which may be sporadic, continuous or a once-only explosion. The result can often be dramatically beautiful coloured nebulae. Planetary nebulae are one manifestation of mass loss and are some of the most pleasing looking nebulae since they often exhibit a satisfying symmetry.

Planetary nebulae

The name planetary nebula was coined by William Herschel to describe the small, circular objects of uniform brightness which he had discovered in his sweeps across the sky in search of nebulae. The appearance of these pale discs reminded him of the planet Uranus; hence their name. Unlike the planets, which have high densities and solid or liquid interiors, planetary nebulae are gaseous bodies. At first, however, it was thought that they were star clusters, like other 'nebulae' which had been resolved into stars with the most powerful telescopes available. Of course at the time neither Herschel nor anyone else had been able to use photography to make any clear distinction between planetary nebulae and clusters of stars or external galaxies. What finally convinced Herschel that some nebulae were indeed gaseous was his discovery on 13 November 1790 of NGC 1514 (Fig. 71) – 'a most singular phenomenon!' (Herschel, 1791). The

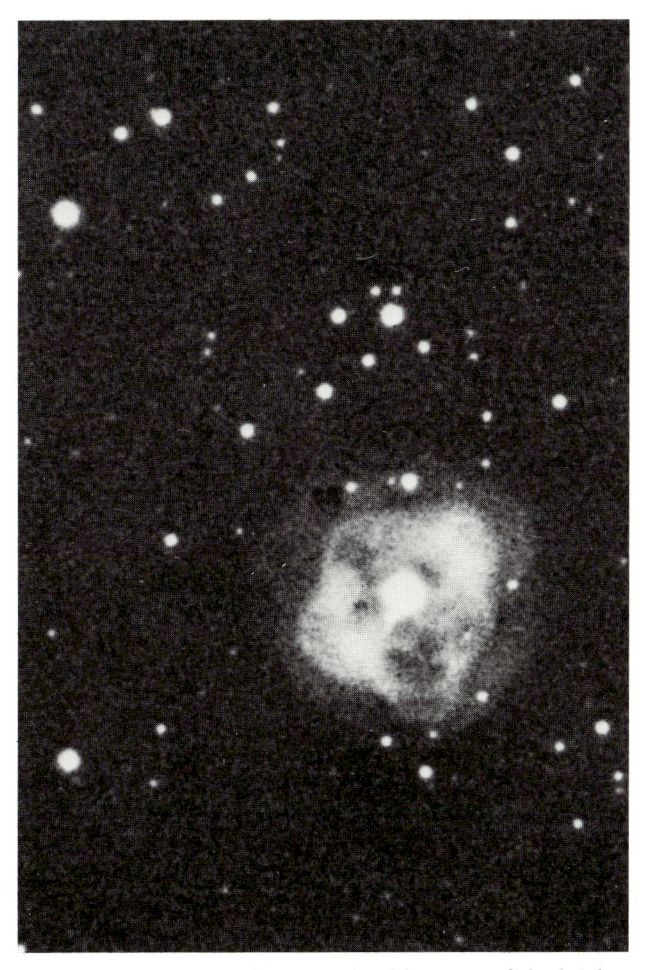

Fig. 71. *NGC 1514. The geometrical location of the bright central star in this nebula convinced Herschel that they were generically connected.*

nebula was 'a star of about the eighth magnitude, with a faint luminous atmosphere, of a circular form, and of about 3′ in diameter. The star is perfectly at the centre and the atmosphere is so diluted, faint and equal throughout, that there can be no surmise of its consisting of stars; nor can there be a doubt of the evident connection between the atmosphere and the star.' Within two months of his observation Herschel reversed his previously held conviction that all nebulae were distant star clusters and admitted the existence of 'true nebulosity', or 'nebulous stars, properly so called'.

The name *planetary nebula* now means any nebula which shows a generally disc-like symmetry, including what Herschel called 'a perforated nebula or Ring of stars' which had a hole in its middle.

However, planetary nebulae defy precise description, a state of affairs which causes concern to astronomers. Dave Hanes has pointed out to us that in the *Proceedings of the International Astronomical Union Symposium Number 34*, a gathering of the world's experts on planetary nebulae, the following dialogue is reported on p. 290:

> David S. Evans
> *It is essential to have a practical definition for a planetary nebula . . . The practical definition now in use seems to be that a planetary nebula is an object which appears in a catalogue of planetary nebulae.*
> Rudolf L. Minkowski
> *. . . there is no better way than to accept any object in a catalogue of planetary nebulae if nobody has serious objections.*

The modern theory of the origin of planetary nebulae is that each results when an elderly star with a mass of 1–1.5 that of the Sun becomes unstable. When the star is in its red-giant phase it will reach a stage in the process where the delicate balance between inward gravitational attraction and outward pressure can no longer be maintained. In a rather short time, about 10% of the star, its whole outer envelope, is thrown off to form the characteristic shell of a planetary nebula. The ejecta expand outwards with velocities of around 20–40 km/s and slowly merge into the interstellar medium. Within a few tens of thousands of years the nebula is too tenuous to be detected. The relatively short lifetime of the distinctive shell accounts for the relative scarcity of planetary nebulae; only about 1000 of these objects are known, although undoubtedly many are undetected throughout the Galaxy. The planetary way of death is thought to be the usual end for many less-massive stars.

Fig. 72. *Nebulium! Five years after the discovery of 'nebulium' H.E. Roscoe reproduced in a book of lectures* On spectrum analysis *the spectral lines as they appeared to Huggins' excited eye for the first time. Hβ (4861 Å) and the two 'nebulium' lines at 4959 and 5007 Å stand out green on a faint continuum.*

The colours of planetary nebulae

To the eye aided by a telescope some planetary nebulae have a greenish blue look. Their spectra reveal why. Unlike stars, planetary nebulae have an emission-line spectrum (p. 3) which shows a few individual colours, not a continuous spectrum exhibiting all colours simultaneously. The most prominent spectral line is the one which appears at 5007 Å. Once known as the nebulium line because it was first found in astronomical nebulae, the spectral line has a green colour (Fig. 72). It was first observed by Sir William Huggins in 1864 when he turned his spectroscope on the planetary nebula NGC 6543. The excitement of his discovery stands out in his account written 25 years later:

> *On the evening of the 29th of August, 1864, I directed the telescope for the first time to a*

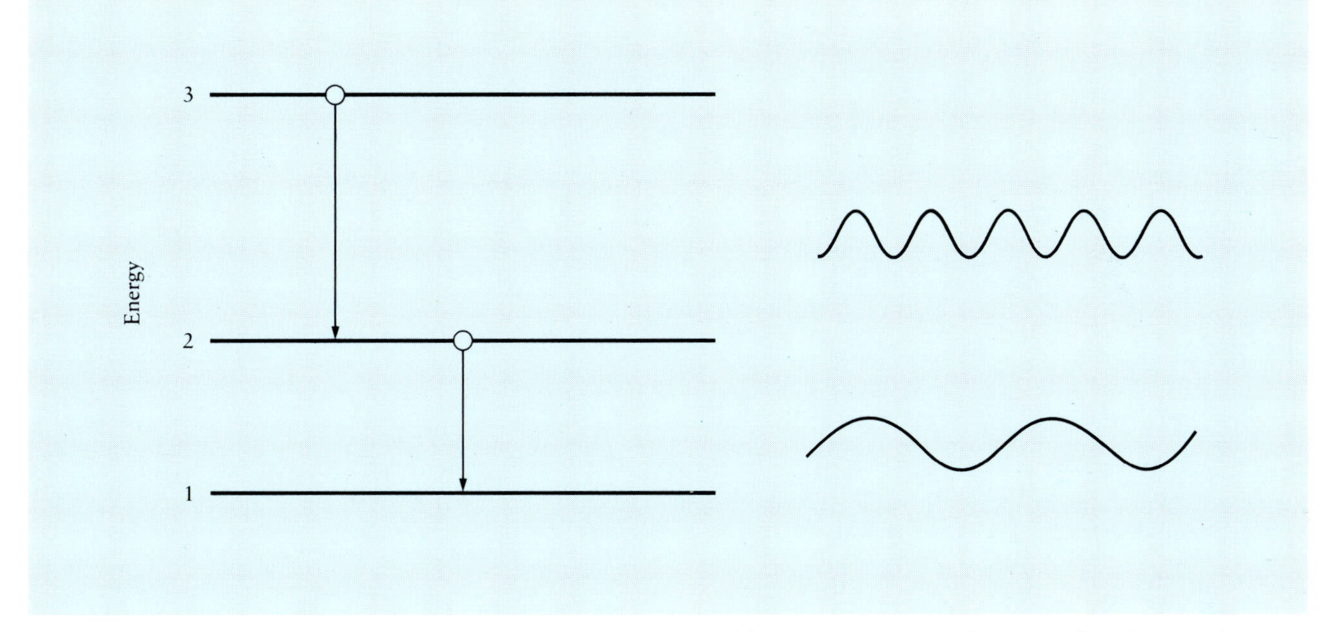

Fig. 73. *Quanta. If an electron drops down a large energy step in an atom it emits a high-energy quantum of short-wavelength radiation. A small energy jump produces a low-energy, long-wavelength quantum.*

> *planetary nebula in Draco. The reader may now be able to picture to himself to some extent the feeling of excited suspense, mingled with a degree of awe, with which, after a few moments of hesitation, I put my eye to the spectroscope. Was I not about to look into a secret place of creation?*
>
> *I looked into the spectroscope. No spectrum such as I expected! A single bright line only! . . . The light of the nebula was monochromatic, and so, unlike any other light I had as yet subjected to prismatic examination, could not be extended out to form a complete spectrum . . . The riddle of the nebulae was solved. The answer, which had come to us in the light itself, read: Not an aggregation of stars, but a luminous gas.*

The line itself, at 5007 Å, and a satellite line at 4959 Å which always accompanied it at one-third the strength, were a mystery. No spectral lines had ever been seen at these places in a spectrum in any terrestrial laboratory (nor still have), unlike, say, the sodium D-lines at 5890 and 5896 Å which are familiar lines in astronomical objects *and* in the laboratory (salt thrown into a hot flame for instance shows the yellow glow). Hence astronomers referred to the source of the 5007/4959 pair as a new element, nebulium. The problem in inventing such a new element was that there were no spaces left in the Periodic Table of elements in chemistry. It had been filled and could not accommodate any unknown element. The riddle was solved in 1927 by I.S. Bowen, following clues by H.N. Russell. Russell suggested that the nebular lines were from atoms of known kinds shining in unfamiliar conditions. He remarked that the nebulium lines may be emitted only in gas of a very low density. To follow his argument we have to look at the way a spectral line is made.

Energy levels in atoms

Electrons orbit around the nucleus of an atom much as spacecraft orbit around the Earth. If a spacecraft receives more energy from forward firing its rocket motors it moves into a higher orbit; if it loses energy by friction with the atmosphere its orbit decays. The radius of its orbit can be changed smoothly from one value to another by adding or taking away the right amount of energy.

Electrons can change orbits by receiving or emitting energy, but these changes do not take place by the smooth processes typical of spacecraft. Instead, electrons can move from one orbital energy to another only by steps. Electrons may not adjust their energy in a way which moves them half-way between these steps – the right energy must be received or emitted, no more, no less.

If a change takes place in an atom so that an electron drops from a higher energy level to a lower level, the energy of the atom decreases and this energy leaves the atom. The loss of energy can take place

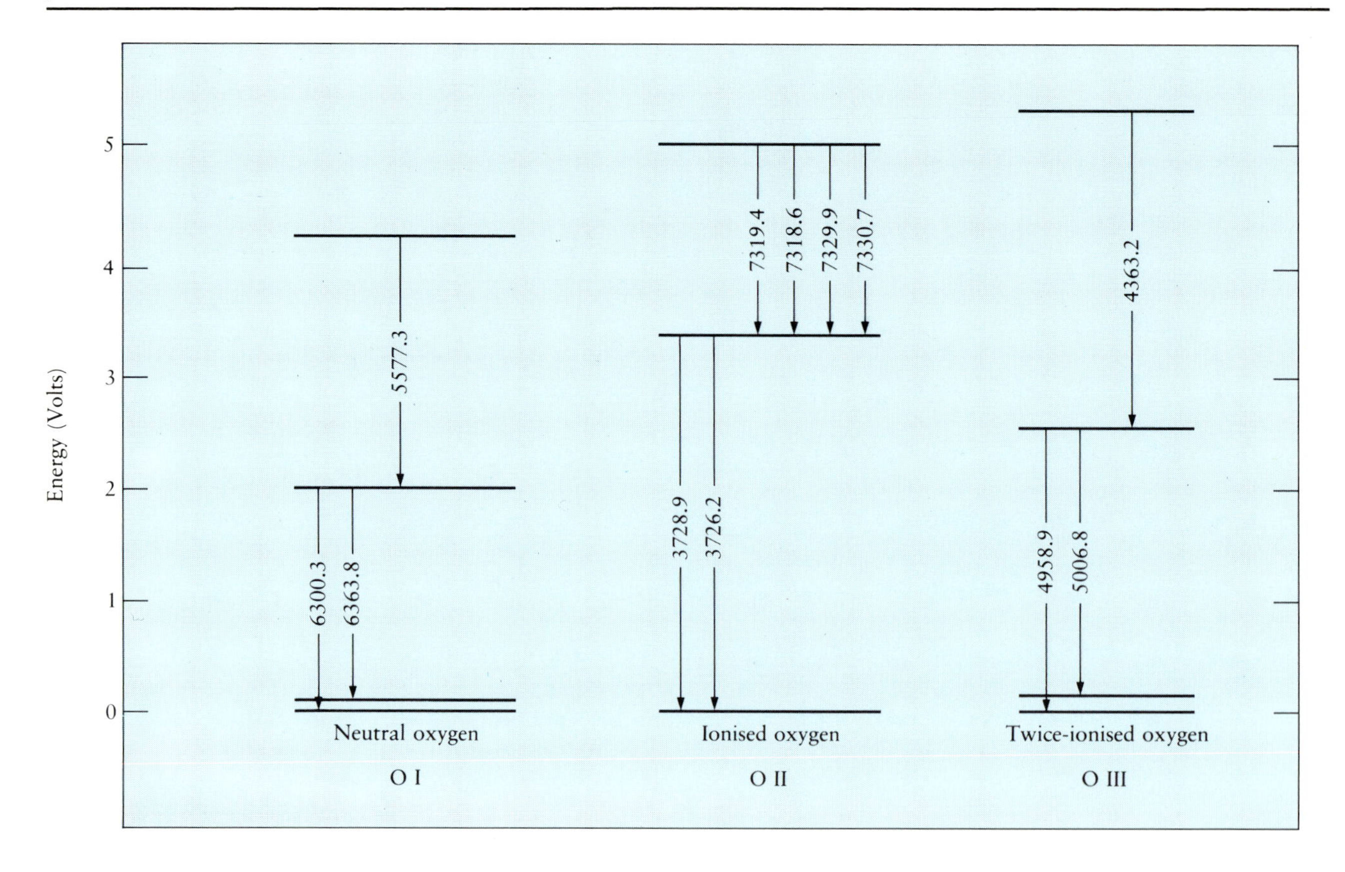

Fig. 74. *Forbidden lines in oxygen. Electrons moving amongst the lowest energy levels of neutral, ionised and twice-ionised oxygen yield the spectral lines whose wavelengths are shown including the 5007 and 4959 Å nebulium lines.*

during a collision between the atom and another, in which case the energy is carried off by the other atom. A second way is for the energy to be converted into light energy in which case a 'packet' of light leaves the atom (Fig. 73). Because the energy levels are in steps (or quantised), the energy of the packet (or quantum) of light which leaves the atom when an electron jumps between two particular levels is always the same. Because there is a relation between the energy of a quantum of light and the light's wavelength, this means that the wavelength of the light from a particular jump is always the same. Hence the colour of the light is always the same, and light from this jump always appears at a particular place in the spectrum.

Now in an atom of a given kind (for instance oxygen) there are many, many energy levels available to each of its electrons. If a collection of oxygen atoms were being stimulated so that electrons in them were passing about amongst all the levels there would be a chaotic jumble of wavelengths of light from the atoms. However, there are rules which restrict the transit of an electron occupying a given level to only a few possible levels, much as the knight on a chessboard can jump to one of only eight squares. Thus the number of wavelengths which a given kind of atom can emit is very restricted. Moreover, some transitions are much more probable than others, so that some wavelengths of light appear more often than others. Thus the spectral signature of a given kind of atom consists often of a few bright spectral lines – an emission line spectrum which identifies the atom as clearly as the pattern of moves made by a chessman identifies the piece.

As in chess, however, there occasionally occur exceptional moves. Normally a king is permitted to move to any of eight adjacent squares. Occasionally, under some circumstances, a king may move two squares ('castling'), although such a move is usually forbidden. Similarly there are *forbidden transitions* in an atom which normally an electron will not make by the emission of a quantum of light, but which may occur in some other circumstances, or which may very *rarely* occur by the emission of a light quantum.

What happens if there is a level in a certain kind of atom from which all transitions resulting in light emission are forbidden? An electron may pass into the level and be unable to get out by means of a permitted transition. In a dense collection of atoms (as in a terrestrial gas), an atom in such a state (called a *metastable* state) will soon be bumped by another atom and be able to pass on the energy to it; no light is emitted, but the electron has made the forbidden transition in another way. If the collection of atoms is very rarefied, as in interstellar space, collisions between atoms seldom occur. Eventually the rare forbidden transitions will take place. In fact they may give rise to the strongest spectral lines visible in the optical spectrum (although permitted transitions may give rise to even stronger spectral lines in the ultraviolet).

I.S. Bowen realised that the two spectral lines at 4959 and 5007 Å in planetary nebulae arose from forbidden transitions from a metastable state of twice-ionised oxygen (Fig. 74). There was no need to invoke the existence of a strange cosmic element, nebulium. The cosmic circumstances of a rarefied nebula in which oxygen atoms had lost two of their electrons were enough to explain the 'nebulium' lines. Other spectral lines were tracked down to 'forbidden' transitions from common atoms or common ions (Table 10).

Behind the green glow of planetary nebulae seen by Herschel and Huggins was the story of a forbidden transition, not a non-existent element! – but what does colour photography show?

The Helix Nebula

One of the finest examples of a planetary nebula is the Helix Nebula, NGC 7293. Our colour picture, made from three plates taken with the AAT (Fig. 75), shows it to be almost half a degree across, about the size of the Full Moon, but we will see later that it is a good deal larger than that. The nebula takes its name from its curious form, reminiscent of what you would see by looking into a few turns of a coiled spring, or down a spiral staircase. On the inner edge of the 'coiled spring' we find numerous red streaks, all

TABLE 10. BRIGHT SPECTRAL LINES OF NEBULAE

Wavelength (Å)	Symbol*	Element	Ion	Forbidden?	Relative Intensity in Planetary nebulae	Relative Intensity in Orion Nebula
ultraviolet						
3425	[Ne V]	Neon	+ + + +	F	30	
3727	[O II]	Oxygen	+	F	30	100
3869	[Ne III]	Neon	+ +	F	50	
3968	[Ne III]	Neon	+ +	F	25	20
blue						
4101	H I	Hydrogen	neutral		25	28
4340	H I	Hydrogen	neutral		40	44
4686	He II	Helium	+		40	
4861	H I	Hydrogen	neutral		100	100
green						
4959	[O III]	Oxygen	+ +	F	300	112
5007	[O III]	Oxygen	+ +	F	800	340
yellow						
5876	He I	Helium	neutral		25	28
red						
6300	[O I]	Oxygen	neutral	F	30	22
6548	[N II]	Nitrogen	+	F	70	15
6563	H I	Hydrogen	neutral		400	300
6584	[N II]	Nitrogen	+	F	150	50

*This notation includes the chemical symbol of the element from which the spectral line arises, and a Roman numeral which is one more than the number of electrons lost by the atoms. Thus 6563 Å comes from H I, meaning hydrogen atoms with no electrons missing. The square brackets distinguish forbidden spectral lines.

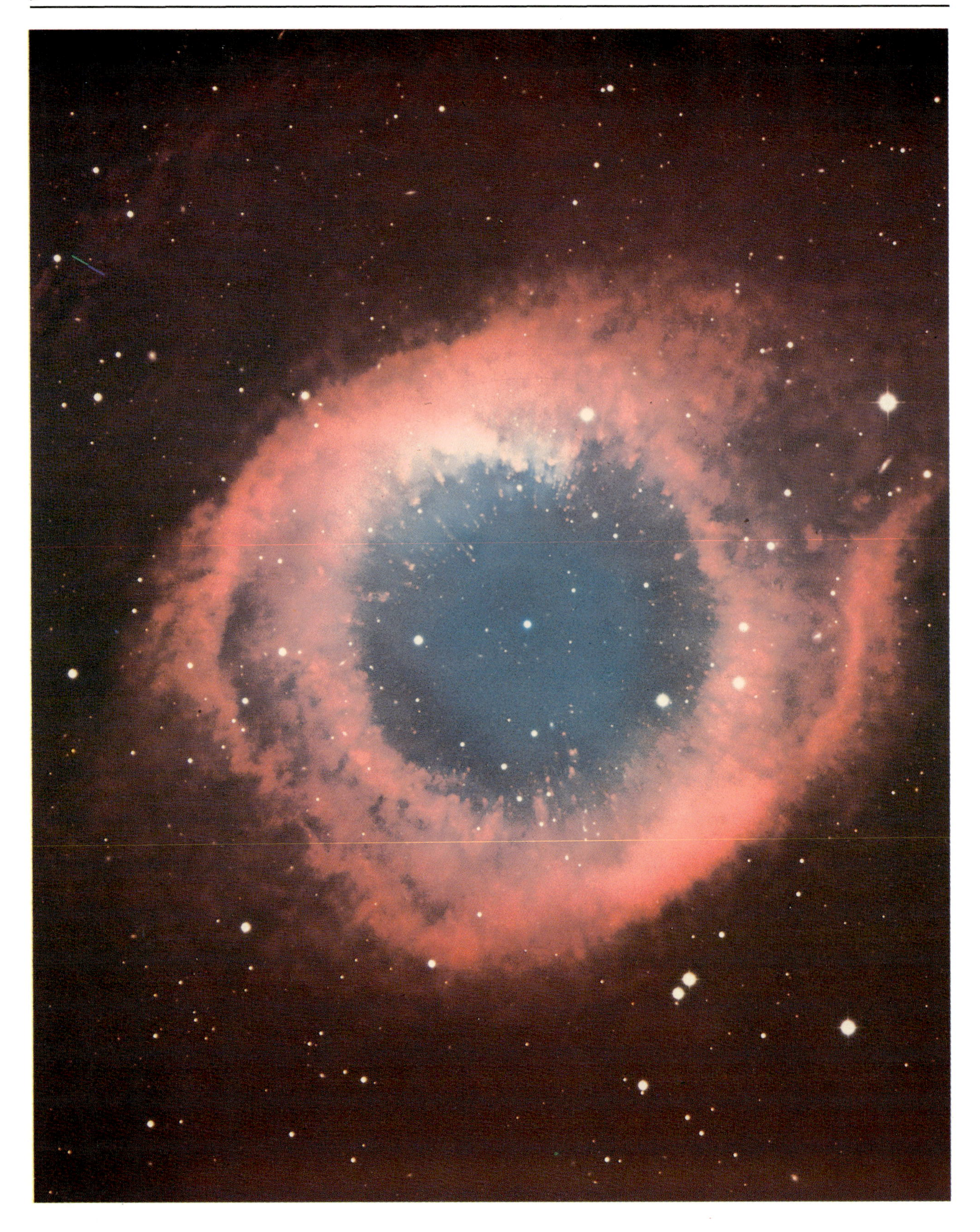

Fig. 75. (left) Helix Nebula. This is the nearest planetary nebula to us, about 500 l.y. away. The colour picture shows the various ionisation levels within the shell of matter ejected from the central star. The smallest radial blobs inside the red shell are thought to be 150 astronomical units (a.u.) across (150 times the Earth–Sun distance) and give this beautiful object its alternative name, the Sunflower Nebula.

Fig. 76. Helix Nebula dissected. The colour separation plates used to make Fig. 75 reveal very interesting differences in morphology when compared. A reproduction of the red plate is printed (a) (right) with an unsharp mask to enhance the abundant fine detail. On the green-sensitive plate, (b) (below), mainly in the light of doubly ionised oxygen, the nebula is almost structureless and the distinctive helix has disappeared.

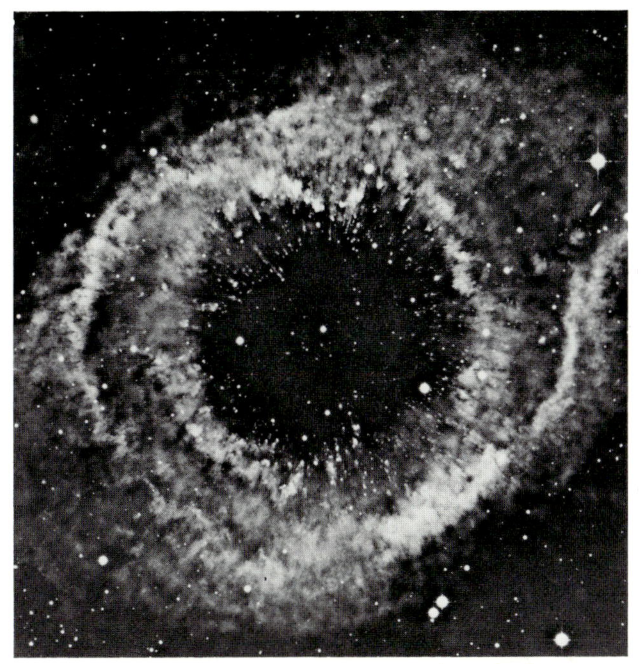

pointing to the central star, confirming that it indeed is the star which exploded. The radial streaks have been likened to the petals of a flower, hence the other name for this object, the Sunflower Nebula. The nature of these streaks is not clear and they are the subject of much speculation. Perhaps they were produced by some subsequent minor explosions on the central star; perhaps they are caused by a steady wind of gas blowing from the central star, overtaking the ejected shell and breaking it into droplets.

The colour picture shows a marked change of colour as we move out from the centre of the Helix Nebula. The central part of the nebula is blue-green, while the extreme outer edge is deep red, which shades pink and white towards the centre. This effect in planetary nebulae is known as stratification. The colours arise from the prominent spectral lines of Table 10 which in turn betray the presence of the ions from which they arise. The ions are created by ultraviolet radiation colliding with atoms and liberating or removing their electrons. Generally the effect of the ultraviolet radiation is more extreme close to the central star; more electrons can be ejected from an atom if the atom is close to the central star than if it is far away. Thus the central part of the nebula shows the ultraviolet and blue spectral lines in Table 10. The outer rim shows the red spectral lines, particularly the hydrogen line at 6563 Å. The mid zone shows the violet 3727 Å line of [O II] and the green spectral lines of [O III] at 4959 and 5007 Å. These are superimposed on the red image and give a pink or white appearance.

The photograph gives a completely different impression of colour from that seen by eye. What Herschel would have called greenish blue appears here as mostly red. Why? Of course the Helix is a faint nebula – and the eye at low intensities sees by the 'rods'; these peak in sensitivity very near the 5007 Å line, but are weak at 6563 Å, Hα, which therefore doesn't register. However, 6563 Å does register on our photograph and dominates the colour.

The stratification of the Helix is confirmed by dissecting the nebula, comparing the individual images used to make the colour superimposition. The red-sensitive plate used for the colour picture appears as the illustration in Fig. 76(*a*). It was made on the AAT in excellent seeing conditions and has been specially printed (via an unsharp mask) to show the fine details. The smallest objects seen are the bright knots at the ends of the radial streaks – these are about 150 astronomical units (A.U.) across (150 times the Earth–Sun distance or a little less than a light day) and several hundred can be seen, together with much

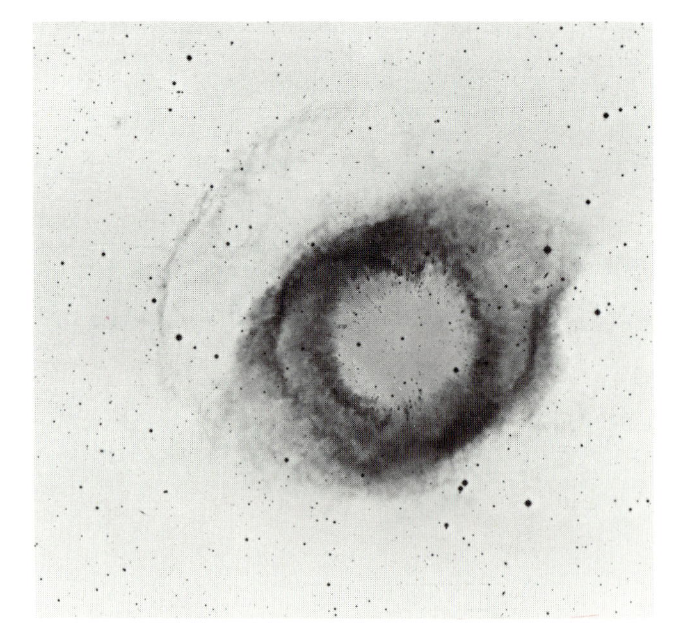

structure in the bright annulus of the nebula. Particularly noticeable are the dark radial streaks crossing the bright annulus. The picture in Fig. 76(*b*) was made from the green-sensitive plate. The lack of fine detail is obvious, as is the quite different shape of the nebula. Gone is the hollow-centered red helix; it coexists with an object which has every appearance of a thick-walled sphere devoid of helical structure.

Photographic amplification of the image on the red plate of the Helix Nebula yields yet more information about this fascinating object. Fig. 77 compares the visual appearance of the original plate with the amplified version. Several new features associated with the nebula are indicated: (A) is a curious cometary object only visible on the red plate while (B) and (C) are structured filaments quite

Fig. 77. *Helix Nebula amplified (below). The red plate used to make Fig. 75 has been subjected to the photographic amplification process to reveal previously unsuspected faint detail unseen in the direct, reproduction (left).*

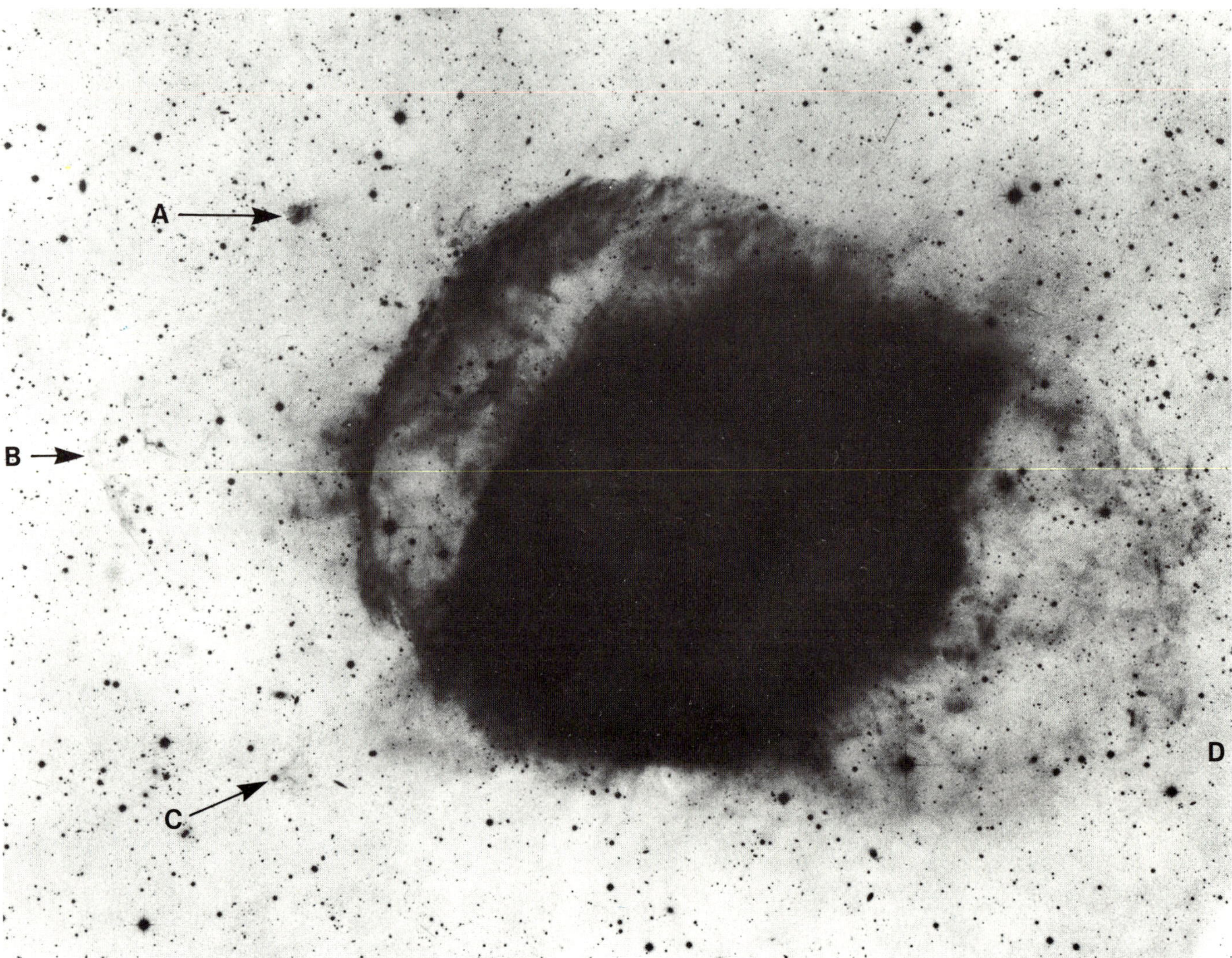

invisible on the original. The loop (D) was discovered by Araya and colleagues (1972) but this is the first photograph to show its full extent. If (A), (B) and (C) can be regarded as constituting a loop, similar to that at (D), then NGC 7293 has a twin-lobed symmetry and is known as a bi-polar nebula.

The shapes of planetary nebulae

What are the reasons for these shapes? Planetary nebulae are not simply the filled or hollow spheres which their name suggests. Some may be doughnut-shaped – toroids, like a car tyre. If a toroid is seen face-on it will appear circular, or annular. Indeed, Herschel considered planetary and annular nebulae as different, although they are now regarded as aspects of the same phenomenon. An annulus could be interpreted as a hollow sphere; but if viewed from the side, it may show as a two-lobed object. At intermediate angles, and depending on the relative thickness of the toroid relative to its diameter, it could have the shape of a cylinder viewed obliquely. Under some circumstances, perhaps if the explosion occurred on a central star which was rapidly rotating, or on one member of a double star, it may happen that the outward flow of gas into the toroid was given a twisting motion. Perhaps this accounts for the helical shape. It seems clear from the different shapes which coexist in the Helix Nebula that the ejection process is not the simple explosive event which we first conjectured, but a more drawn-out process, or series of processes, in which the outflow of matter is modified by circumstances and carries traces of their history. Presumably the faint outer extensions seen in several planetary nebulae represent previous explosions, or matter expelled from the parent star while it was a red giant and before it produced a planetary nebula.

An example where the butterfly shape of a bi-polar nebula is clearly developed is NGC 6302 (Fig. 78). This object is an unusual planetary nebula because velocities of around 400 km/s have been found in the ejected gas. This is higher by a factor of ten than the velocities commonly found in planetary nebulae and suggests that the initial explosion was a particularly energetic event. The reason for the curious butterfly shape may be that rotational or magnetic forces were involved in directing the expansion of the ejected material. It was funnelled into two opposite directions, like a long balloon bursting simultaneously at both ends. Perhaps the axis so defined is the rotation axis or magnetic axis of the star.

Fig. 78. *NGC 6302. This is one of a large number of planetary nebulae with bi-polar symmetry showing a distinctive bow-tie or butterfly shape.*

Stars in outburst

There is no straightforward definition of a planetary nebula and it is therefore sometimes difficult to classify objects which show some or even most of the necessary properties and yet deviate in some way from an ill-defined standard. NGC 6164/5 is one such. It is clear from Fig. 79 that the bright central star is surrounded by a pair of well-defined shells and other, more diffuse nebulosity. (The NGC numbers refer to these shells as separate objects.) The central star, known only by its catalogue name HD 148937, is part of the classification problem. Planetary nebulae usually have rather faint stars at their centres, the dying inner cores of red giants which quite rapidly evolve into white dwarfs. These stars are small – typically Earth-sized – and have about the same mass as the Sun. A star of this kind is the power source of the Helix Nebula. NGC 6164/5 is ten times more distant than the Helix yet its central star appears 40 times brighter. It cannot therefore be a white dwarf. In fact HD 148937 is the most prominent member of a triple star system which seems to be losing mass in an irregular way and has been doing so for several tens of thousands of years.

An earlier outburst, now fading away into interstellar space, can be seen in Fig. 80. The billowed structure is due to its expansion into the non-uniform interestellar medium.

It is unusual for young stars like HD 148937 to eject sufficient material to produce a visible nebula – unusual, but not unique. Another class of objects, to which HD 148937 seems to be quite closely related, are the Wolf–Rayet stars, named after their French

Fig. 79. *(above)* *NGC 6164/5. Superficially similar to a planetary nebula, this is in fact a rather unusual object. At its centre, a young blue, very hot star, the brightest of a triple system, is losing mass continuously. Occasionally a much more vigorous outburst gives rise to the shells and nebulosity seen here. It is not clear whether the ejected material is thrown off at the equator of the star, giving rise to a tyre-like circumstellar ring, or at the poles of the star, giving rise to two jets.*

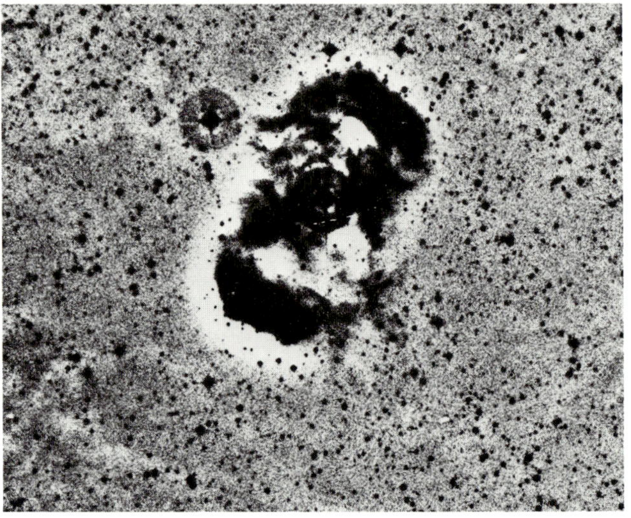

Fig. 80. *(right)* *NGC 6164/5 amplified. Evidence of an earlier outburst of the peculiar star HD 148937 at the centre of NGC 6164/5 is visible in this amplified photograph. This outer region can only be shown clearly if the background nebulosity is removed photographically, a process which gives rise to the spurious white halo around NGC 6164/5. The brightest part of the outer nebula in the south-east corner seems to be colliding with another nebula, NGC 6188, which is mostly off the print, but wisps from this can be seen in the lower left corner.*

Fig. 81. *NGC 2359. Pictured here is a Wolf–Rayet star inhabiting part of our Galaxy which is rich in hydrogen gas. A vigorous stellar wind from a hot blue star has created a cosmic bubble of highly ionised gas. Evidence of prolonged activity of this kind is all around in the red streamers of photo-ionised hydrogen which cover a large expanse of sky.*

discoverers. These stars, too, are often multiples, but their main feature is their very vigorous stellar wind which accelerates surrounding material to enormous velocities. The winds are so violent that a visible shock front is produced, an interestellar bubble of ionised gas surrounding a very hot star. NGC 2359 is a fine example of this rare class of object. The colour picture (Fig. 81) shows that the nebulosity associated with this star is quite extensive and suggests a long history of outbursts in a particularly dense region of interstellar matter.

Supernova remnants

The most extreme kind of stellar explosion is called a supernova explosion. At the end of the life of a massive red giant star its energy source dies away and the flow of energy outwards in the star drops below the critical level needed to balance its gravitational force. No longer is the body of the star supported against its fall and the star collapses. There is a huge release of gravitational energy in the central regions of the star. The gravitational energy is released as a flood of radiation, which streams outwards and turns around the inward falling outer parts, transforming the sudden collapse into a gigantic explosion.

This catastrophe occurs in a matter of minutes. The consequences of the explosion persist long after. First, the explosion is seen as the flaring up

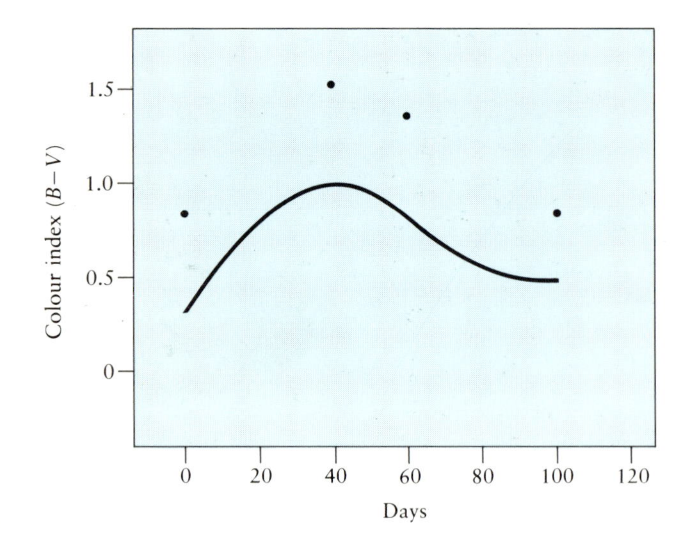

Fig. 82. *Colours of Tycho's supernova. Observed points lie above the usual supernova colour curve, indicating that Tycho's supernova was redder than normal.*

TABLE 11. COLOUR OF TYCHO'S SUPERNOVA

Date	Colour	$(B-V)$	$(B-V)_0^*$	$E(B-V)^\dagger$
Nov. 1572	Like Venus and Jupiter	0.83 / 0.82	0.3	0.5
Jan. 1573	Like Mars and Aldebaran	1.36 / 1.52	1.0	0.5
April 1573	Similar to Saturn	1.04	0.5	0.5

* $(B-V)_0$= the colour index at the equivalent age of supernovae (of a similar type to Tycho's) when they are seen through no dust.
† $E(B-V)$= the colour excess of Tycho's supernova and is the difference between $B-V$ and $(B-V)_0$. It represents the amount of interstellar reddening towards Tycho's supernova.

of a previously unnoticed star; it is this appearance of a bright seemingly-new star which is called a supernova. The supernova reaches its maximum brightness within a day or so, as the heated outer parts of the star rapidly grow in size in the explosion. For a matter of weeks the supernova may outshine the totality of all the other stars in the same galaxy. Further increase in size of the expanding star causes the surface of the exploding shell to cool, so that over a few years the supernova gradually fades. The progressive cooling causes supernovae to redden at first. Other effects in their spectra cause them to become bluer after about 40 days (Fig. 82).

Supernovae occur about once per generation in a given galaxy – say once every 30 years. However, they tend to occur in the dusty parts of the Galaxy and are often dimmed into invisibility – in fact the last supernova seen in our Galaxy was in 1604. The one before that was in 1572. They are known by the names of two great astronomers who observed them extensively, Johannes Kepler and Tycho Brahe, respectively, although these are but two of a list of astronomers known to have seen the supernovae. Because supernovae are rare, the observations made even in that pre-telescopic era have been collected and analysed for every morsel of information. This includes the colour of the supernovae.

Some oriental records give directly the colours of the supernovae. For instance, on the nineteenth day that Tycho's supernova was visible (24 November 1572), when it was about as bright as Jupiter, the Chinese astronomer who wrote the annals of the Emperor Shen-tsung said that 'the star was orange in colour. It was as large as a lamp.' European astronomers likened it to Venus and to Jupiter in colour. By January 1573, when it had faded relative to Jupiter, it had reddened and was similar to Mars and Aldebaran. In April 1573 it was compared in colour to Saturn (Table 11). These observations are consistent with the known progression of colour of supernovae as they fade (Fig. 82). Note however that all the observations of Tycho's supernova lie above the curve observed for supernovae in other galaxies. Tycho's supernova looked redder than these. The shift represents the interstellar reddening towards the supernova, which (Table 11) has a value of half a magnitude of colour excess. Using these archival observations Sidney van den Bergh (1970) has calculated the intrinsic brightness of Tycho's supernova and related it to the size of the expanding Universe deducing a value of the Hubble constant of about 100 km/s/Mpc (see Chapter 8). He remarks that 'this estimate, based on ancient observations of a single supernova is of course highly uncertain' but it is noteworthy that any meaningful conclusions can be drawn at all!

As the ejecta from the supernova speed at many thousands of kilometres per second into space, they collide with the interstellar gas nearby. In the shock of the collision, the energy released stimulates the gas to glow, and beautiful nebulae are produced. Fig. 83 shows the Vela supernova remnant, the

Fig. 83. *Vela supernova remnant. Red and blue filaments are visible, part of the 6° of the whole supernova remnant. Generally the blue filaments lie outside the curve of the red.*

remains of an unrecorded supernova occurring 10 000 years before the Christian era, and visible over 6° of the southern constellation of Vela. Optical astronomers can see only half of the remnant, the other half lying behind a cloud of interstellar dust. Fig. 83 shows only a small part of what can be seen. The remnant has a lacy appearance, with the material broken into filaments of red and blue. The red colour comes from the Hα emission at 6563 Å; the blue colour comes from the [OII] line at 3727 Å. The picture shows the blue emission often lying outside the red; this is a stratification effect showing the heating of the interstallar gas in the shock of the explosion. The hotter material emits the 3727 Å line more than does the cooler material. The stratification of the Vela supernova remnant is the reverse of that of the Helix planetary nebula, because the hotter part is on the outside of the nebula where the collision between ejected material and the interstellar medium is strongest.

The blue parts of the nebula can only be seen to the east of the great shell of the supernova remnant. On the western side, the blue is suppressed and the red is weakened by a curtain of interstellar reddening half drawn across the supernova remnant. Although this curtain half obscures the view, it provides the only credible estimate of the distance to the supernova remnant. Astronomers probe the curtain of obscuration by observing stars at greater and greater distances, until they find ones which show the effects of the interstellar reddening. This gives the distance of the curtain, and the supernova remnant lies just behind. This weak evidence gives a distance for the Vela supernova remnant of 500 pc (about 1500 l.y.). This datum is an important straw to clutch at, because inside the Vela supernova remnant lies the so-called Vela pulsar, the stellar cinder formed in the Vela supernova explosion. A pulsar is a radio star showing a regular and rapid radio pulse as a radio beacon on the star sweeps its beam across the Earth, like a radar antenna. The Vela pulsar is one of the two pulsars which also show optical emission – the other lies in the Crab Nebula. The Vela pulsar is very much fainter than the Crab pulsar, the reason being connected with the rapid fading of the optical beam as the pulsar ages. To get an idea of the rate of fade, and therefore of the fading mechanism, it is important to theoretical astronomers to be able to compensate the observed brightness of the Vela pulsar for the effects of distance – and of course this can only be done if its distance is known. The curtain of interstellar reddening, half obscuring the supernova remnant, provides the only indication of the pulsar's distance and the barest check on theoreticians' ideas of the pulsar aging mechanism.

There are some supernova remnants where we see not the collision between the exploding star and the general interstellar material lying around it but the collision between the exploding star and material ejected by the star during a previous stage of its existence. For instance a Wolf–Rayet star like HD 148937 may eject a nebula like NGC 2359 and then at a later time, when it has exhausted all its nuclear fuel, it may explode as a supernova. The clues to this process are in the spectrum of the supernova remnant: it shows not the typical spectrum of interstellar material with abundant hydrogen but an abnormal spectrum enriched with other elements.

An example of this is the supernova remnant Puppis-A, a strong radio source near the line of sight to the Vela supernova remnant. Associated with a shell-like radio source is a motley collection of filaments which reveal themselves photographically only on red-sensitive emulsion. They emit, not the usual strong Hα line at 6563Å, but the the two nitrogen lines at 6548 and 6584Å. The material of the Puppis-A supernova remnant is thus nitrogen-enriched. This is typical of material in the atmospheres of Wolf–Rayet stars, with nitrogen being one of the elements manufactured in massive stars and released in their various kinds of stellar winds.

Unfortunately, because of the understandable lack of interest by astronomers in taking pictures with nothing on them, we have no blue or green pictures of Puppis-A in the plate archives of the Anglo-Australian Observatory, and cannot make a three-colour picture, which would in any case only show red nebulae and more or less white stars. We have only a picture taken in red light which we show on this page in a black-and-white version. The curlicues of gas on the picture represent material leaked off the Puppis-A precursor star and lit up as the supernova explosion overtakes the matter lost a star-generation ago.

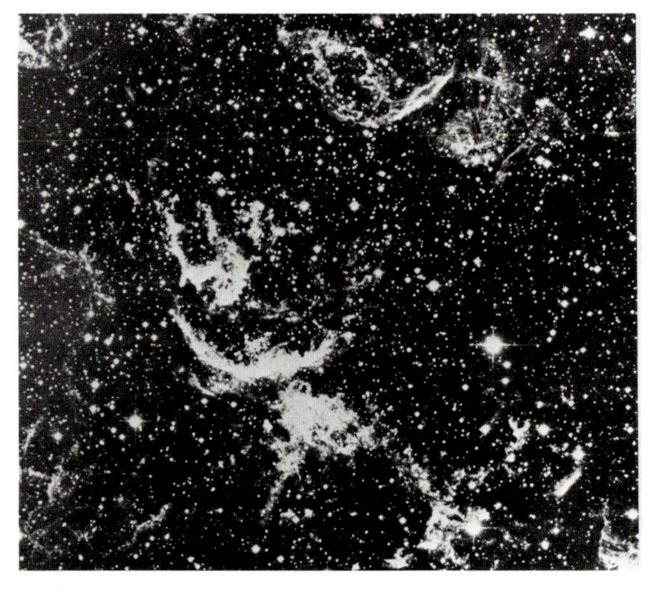

Puppis-A. Curlicues of nebulosity, shining by red nitrogen emission, define matter ejected in a stellar wind from the Puppis-A precursor star before it exploded as a supernova some 4000 years ago.

Chapter Six

Stars from nebulae

The birth of stars from nebulae

In Chapter 5 we saw how stars at the end of their lives can eject material and produce beautifully coloured nebulae. In this, stars are contributing to a cyclic process, returning matter to the interstellar medium from which they themselves were made. For it is amongst nebulae that stars are born and it is because young stars are present that the nebulae can be seen.

There are probably several kinds of events which can trigger the birth of stars. If a supernova explodes near an interstellar cloud, the blast wave from it may compress the cloud and so make it more dense. Two galaxies may pass close to each other and each compress the interstellar clouds in the other. Two interstellar clouds may collide and coalesce. It seems that there are many processes which can set off a wave of compression (density wave) in the interstellar medium of a galaxy.

Whatever the initial cause of the compression of an interstellar cloud, the cloud contracts and becomes denser, and stars form. The collapse of the cloud takes place rapidly and many stars may be formed at much the same time and, clearly, from the same material. It is this homogeneity which makes the study of star clusters such a fruitful one.

The formation of stars takes place by the fragmentation of the interstellar cloud in its collapse, and it is the fragmentation process which produces the spread of stellar sizes, with a relatively large number of small stars and few large ones. As we saw in Chapter 4, these stars form a Main Sequence of colour and brightness stretching from cool, faint red stars to hot, bright blue ones, and it is the latter, shining into the parts of the gas cloud left over, which produce the bright emission nebulae known as H II regions (pronounced 'H-two regions').

Density waves propagate in a spiral around the parent galaxy causing interstellar clouds to collapse, making new stars and H II regions in sequence. Viewed from outside the galaxy, the new stars, the nebulae and the dust clouds comprise the brightest parts of a spiral arm. The curved bands of star formation along the arms are clearly shown in the colour picture of the southern spiral galaxy NGC 2997 (Fig. 125).

The typical density of hydrogen in interstellar space near the plane of our Galaxy is one hydrogen atom per 10 cubic centimetres (cm^3). (Imagine a fairground balloon full of hydrogen expanded to the size of a small planet; the density of hydrogen inside is then comparable to the density of hydrogen in interstellar space.)

An interstellar cloud has a density 100 times this typical value (about 10 atoms of hydrogen per cm^3). After the cloud has been compressed and formed new stars, and become an H II region, its density may be a million times yet greater, still a good vacuum by terrestrial standards.

H II regions

The Victorian spectroscopists listed, in order, the various spectra which appeared from each element as it was excited. After the structure of the atoms was understood, it turned out that I referred to the neutral atom, II to an atom which had lost one electron (a positive ion), III to an atom which had lost two electrons (an ion with two positive charges), etc. The terminology has stuck, so H II refers to an ionised hydrogen atom (H^+). It is appropriate to define the cause of ionisation in a hydrogen atom a little more carefully. As we saw in Chapter 5, there is a series of energy steps, or quantum levels, in an atom, at which an electron can sit. The lowest step, the base level, is called the *ground state*, for obvious reasons. An electron can be bumped up the ladder of energy levels, to steps above the ground state, but will eventually fall back down to the ground state (with the emission of light quanta). In the vastness and emptiness of interstellar space, the bumping together of atoms is so infrequent that generally all hydrogen atoms are in the ground state. If, however, there is an energetic collision between two atoms, or between an atom and a random free electron, or, more likely, if an energetic photon passes near enough to the atom, the ground-state electron may be bumped so far up the ladder of energy levels that it becomes, itself, free – the electron has been given so much energy that it has leapt from the hydrogen atom and the atom is now ionised. The amount of energy which is necessary to achieve this is the same as in an ultraviolet photon of wavelength 912 Å. A photon with this wavelength, or shorter, has the energy to ionise hydrogen.

If there is a source of such photons in a cloud of hydrogen, then nearby hydrogen will be ionised. Such photons are made by hot stars – stars whose spectral type is O in fact. Only O-type (and B0 and B1) stars radiate sufficient ultraviolet to be the *exciting stars* of an H II region.

Fig. 84. *M42. The wide range of surface brightness in the Orion Nebula usually means that the inner regions are burned-out, even on short exposures. Three 5-min exposures were combined to make this picture; they were first copied through an unsharp mask to retain detail throughout the nebula without distorting the colour balance. The northernmost nebula, seemingly detached from M42 in this light exposure, is M43. Subsequent more deeply exposed pictures show it as part of M42. Interstellar reddening by the absorbing cloud which divides the two nebulae produces the orange glow around the black patches over the bright parts of the nebula north-east (upper left) of the Trapezium stars.*

Fig. 85. *Trapezium. Positive copies of the plates used to make Fig. 84 were enlarged to produce this photograph of the Trapezium stars at the heart of the Orion Nebula. These young, very hot stars produce much of the energetic ultraviolet radiation which causes the nebula to glow in the characteristic green light of oxygen and red of hydrogen. These colours combined give rise to the yellow coloration of the central part of the Orion Nebula. The lack of faint stars in the upper left corner of the photograph shows the existence of interstellar absorption in a cloud which curls over in front of the H II region and reddens the upper left border of M 42 and the light which comes around the edges of the dark patches north-east of the Trapezium.*

New, massive stars immediately ionise any surrounding region of hydrogen. As the hydrogen ions and free electrons begin to recombine, an emission nebula switches on to announce the birth of the hot stars. The colours which mark such nebulae are those of the spectral lines which are produced as the hydrogen and other ions recombine (Table 10). The most intense spectral lines are Hα (6563 Å) from recombining hydrogen, and the 5007/4959 nebulium lines.

Great care has been taken to ensure that the colour reproduced in the accompanying pictures is what you would see if you could somehow increase the colour sensitivity of your eyes. The three plates from which the additive pictures were made were taken with a combination of emulsions and filters which give a uniform spectral response over the whole visual spectrum. Colour films on the other hand have some quite large blind spots in their spectral coverage. Of little consequence for normal terrestrial photography, these gaps distort the colour rendering of objects which emit most of their light at a few discrete wavelengths (see Chapter 2). In the case of some nebulae, much of their radiation comes from the 'nebulium' lines at 5007 and 4959 Å, where colour film has a very poor response, but this characteristic green light can be caught with the three-colour addition process. Colour pictures on colour film generally show the nebulae as predominantly red because colour films are highly sensitive to Hα emission.

The Orion Nebula

The brightest nebula visible from Earth is the Orion Nebula, surrounding the star Theta-1 Orionis, the central star in the Sword of Orion. Theta-1 is in fact a cluster of four stars known as the Trapezium. The Orion Nebula, also known as the Great Nebula, was

Fig. 86. *(left) Orion Nebula. Three masked photographs from the UK Schmidt Telescope have been combined in this colour picture of the Orion Nebula. Traces of the dust cloud in which the nebula is embedded can be seen around the nebula, obscuring faint stars beyond. Pink shades from combined Hα and [O III] 5007/4959 Å lines pervade the central part of the Orion Nebula, arising from the H II region excited by the Trapezium stars – the pink area may be visualised as the inner surface of a hollow bite from an apple, which has exposed the seeds (the Trapezium) within. The brightest star at the bottom of the picture is Iota Orionis surrounded by haloes arising in the telescope optics. At the top of the picture is NGC 1977. Its predominantly blue colour marks it as a reflection nebula. The same colour surrounds M 42 and comes from starlight reflected by the dust grains.*

Fig. 87. *Trapezium cluster. The Trapezium stars are part of a small cluster, most of whose members are hidden beyond the Trapezium, in the dusty interior of the Orion Nebula. Infrared radiation penetrates dust more effectively than shorter, visual wavelengths so that a photograph in blue light* (a) *(below) shows little other than the swirling gas and dust behind the Trapezium, whereas the infrared picture* (b) *(right) reveals many of the obscured background stars.*

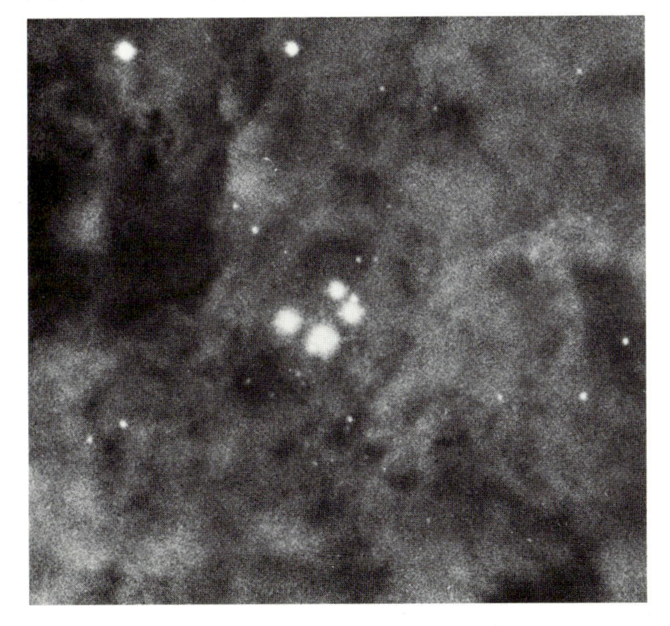

discovered by N. Peiresc in 1610 with a telescope given him by his friend Galileo. It can be seen with the unaided eye as a misty patch surrounding Theta-1, and certainly with binoculars, or even the smallest telescope. Messier catalogued the Nebula as M42 and it was the first extended object to be photographed, by Henry Draper in 1880 (Chapter 2). A more recent picture appears in Fig. 84.

The bright central parts glow with a mixture of red and green (i.e. yellow) emission as shown here, and are surrounded by a fainter but more extensive red component due to ionised hydrogen. The strong lines of [OIII] (4959/5007 Å) and Hβ (4861 Å) in the inner regions of the nebula are just bright enough to activate the green-sensitive cones in the retina when seen in a good telescope, so some observers report seeing the region around the Trapezium stars as greenish, but we ourselves have never seen the Orion Nebula as coloured at all, merely the pearly gray of dark-adapted rod vision.

Because of its proximity, size and brightness, M42 has been extensively studied at all accessible wavelengths and as a result more is known about it than about any other nebula. Essentially, what we see is a complex, inhomogeneous mixture of gas and dust and a scattering of bright stars, some wreathed in nebulosity. At the heart of the complex lies the Trapezium cluster, easily seen with a modest telescope and visible as a group of four stars in the centre of Fig. 85. The Trapezium seems to be the powerhouse of the Orion Nebula. These young stars are probably no more than 100 000 years old and have blown a bubble in the gas and dust involved in their birth allowing energetic ultraviolet radiation to penetrate even further into the surrounding cloud. To the east (left in these photographs) some remaining dust curls in front of the hemisphere blown clear by the stars. Because the tiny particles of interstellar dust are so efficient at scattering blue light, the nebulosity behind appears yellow or even reddish (Fig. 86).

To understand the geometric picture, think of the whole nebula – M42, M43 and NGC 1977 – as an apple of gas surrounded by a thick skin of dust which hides the background stars (Fig. 86). As Louisa Murdin has pointed out, a large deep bite into the near side of the apple reveals the pips inside – the Trapezium stars. The surface of the bite shows as the Orion Nebula, the H II region ionised by the stars within. The stars to the north occupy a position akin to the stalk of the apple and ionise gas surrounding them, like leaves. Their light reflects from the dust in the skin.

Fig. 88 *Orion Nebula. On deep pictures, the whole Orion Nebula region is filled with tangled bright nebulosity and dust clouds. If the deep plate* (a) *(top), taken mainly with Hα light, is printed with an unsharp mask (b) (above), it shows the internal structure of the nebula, and the boundaries between parts of the nebula and its surrounding cloud.*

The Trapezium group are the most obvious members of a small cluster of stars almost hidden by interstellar matter. This obscuration and the brightness of the surrounding nebulosity makes them very hard to see, and indeed most of them cannot be seen at all at visual wavelengths. The two pictures in Fig. 87 were taken in blue light (*a*) and in the near infra-red (*b*) and are both very short exposures. The infra-red picture shows the faint background cluster. Many of these stars are still forming from the dusty gas cloud which envelops them. One day they may be hot and bright enough to burn through their heavy veil and could rival the brilliance of the Trapezium stars.

The Orion Nebula is the brightest region of a much larger nebula. Fig. 88 (*a*) is the deepest wide-field picture of Orion Nebula ever published and shows the extremely complex structure of a part of the surrounding cloud. In Fig. 88(*b*) this same plate has been printed through an unsharp mask to display the tangled web of nebulosity in the outskirts of M42.

The Horsehead Nebula

A few degrees to the north of M42, the same cloud is illuminated by the bright stars Sigma and Zeta Orionis to produce what is probably the most famous image in astronomy, the Horsehead Nebula. The Horsehead is really too faint for successful photography with colour film, so a new version (Fig. 89) has been made by adding black and white plates from the UK Schmidt Telescope.

The Horsehead is an excellent example of a dark nebula, only made visible by its encounter with

Fig. 89. *(right) Horsehead Nebula. This beautiful nebula is part of the same gigantic cloud of hydrogen as the much brighter Orion Nebula. The Horsehead itself, probably the best known image in astronomy, is a dark dust cloud penetrating a bright red strip of photo-ionised hydrogen, IC 434, energised by light from Sigma Orionis, the bright star on the right of this picture. Zeta Orionis emits less ultraviolet radiation, but is the brightest star in visible light in this picture, and its light is reflected from dust nearby to give the blue tinge north of IC 434. A dusty H II region, NGC 2024, lies east (left) of Zeta Orionis. Other small, blue reflection nebulae in this picture are IC 432 and 431 (top edge, left and right of centre), IC 435 (left edge of photograph), and NGC 2023 (above and left of Horsehead). Notice the bright rim of light from IC 434 around the outline of the Horsehead. If the Horsehead were simply a foreground object projected against IC 434, it is unlikely that IC 434 would have any brightening, so exactly fitting the outline. This proves that the Horsehead actually penetrates IC 434. Notice too the lack of stars on the left half of the picture, showing that the Horsehead is merely the most prominent part of a larger dark cloud.*

IC 434, a long bright strip of red Hα emission running approximately north–south. The brightest rim shows evidence of several smaller intrusions of dust between the Horsehead and Zeta Orionis at its northern end. However, most of the energy required to ionise the hydrogen of IC 434 comes from Sigma Orionis; although apparently fainter, this star emits more ultraviolet light than Zeta Orionis because it is hotter.

The opacity and extent of the dark cloud can be judged by counting the relative number of stars on either side of the bright rim which defines its edge. The dark cloud is not quite dark however. Stars embedded inside produce the blue nebula on the dark side of the Horsehead. Reflection nebulae are the subject of Chapter 7.

Barnard's Loop

Barnard's Loop surrounds the whole of the Orion Nebula/Horsehead complex. Fig. 90 is a photographically amplified print made from a plate of this region taken in red light with a 4× 5 inch view camera mounted on the Australian National University's 16-inch telescope at Siding Spring. The original exposure was on hypersensitised red-sensitive IIIaF emulsion exposed through a Schott RG630 (red) filter for a little over an hour at f/3.3. The Loop was discovered by Barnard in the course of his survey of the Milky Way referred to in Chapter 3. Its light consists not only of Hα light from ionised hydrogen but also of light from the stars within Orion, which shows up particularly in photographs obtained from space with ultraviolet light (it was one of the first objects to be photographed from space, in 1966, by a small camera carried in Gemini XI by astronauts C. Conrad and R. Gordon). The Loop has been described as a cosmic bubble blown in the interstellar gas and dust by radiation pressure from the energetic stars in the Orion nebula. Whatever its origins this rather faint object remains a challenge for astrophotographers. To the north of Barnard's Loop is the Lambda Orionis Nebula, a smaller and probably younger version of the loop. This almost spherical shell of gas has been heated by young stars inside it and is expanding rapidly into the massive dark cloud of hydrogen which fills much of the constellation of Orion.

Eta Carinae and its nebulae

The Eta Carinae Nebula is a jumble of nebulae, clusters and bright stars seen in perspective along the Carina spiral arm of our Galaxy. The wide-field picture from the UK Schmidt (Fig. 91) shows the red colour typical of H II regions, split by three broad lanes of dust in a Y-shape. The inner white-looking areas are superimpositions of the strong red Hα, green 4959/5007 [O III] lines and blue 3727 [O II] lines. There is sufficient radiation in the heart of the nebula to activate equally all three colours in the three-colour process; these areas are also rich in associations of hot bright O-type stars which also tend to saturate the photograph and give it a white appearance. The close-up photograph (Fig. 92) shows the O-stars in groups; measuring the brightness of these stars shows the Eta Carinae Nebula to lie at a distance of 6800 l.y.

Eta Carinae itself is the brightest star at the centre of Fig. 92, adjacent to the brightest part of the nebula, in which lies a dense dark patch known as the

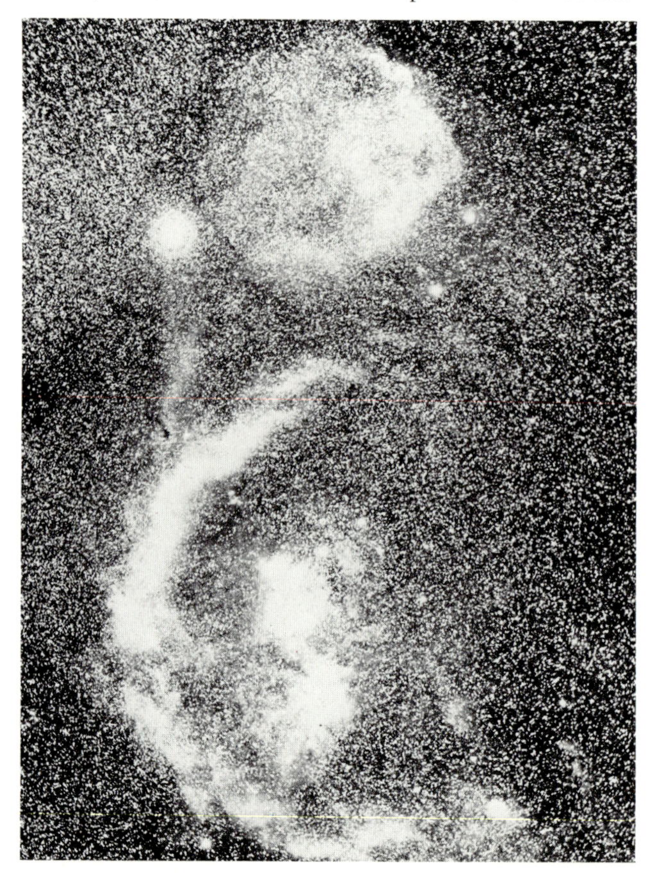

Fig. 90. *(above)* *Barnard's Loop. This very large but faint loop of nebulosity was discovered in 1895 by the pioneer astronomical photographer E.E. Barnard. The ellipse of nebulosity partially surrounds two bright nebulae – NGC 1942 (the Orion Nebula) and, above it, the Horsehead Nebula, which are clearly part of the same cloud. To the north of Barnard's Loop is the Lambda Orionis Nebula, a younger version of the loop. A curious 'spur' or 'wake' runs north from Barnard's Loop and appears to end at the brightest star on the photograph, Betelgeuse. No-one knows if this is a coincidence or whether Betelgeuse is related to Barnard's Loop in some way.*

Fig. 91. *(right)* *Eta Carinae Nebula. The Eta Carinae Nebula lies in a bright part of the southern Milky Way and is thus always below the horizon for observers north of 30° latitude. It is just visible to the unaided eye, and a small telescope or binoculars shows the three dust lanes clearly.*

Keyhole. Fig. 93, an enlargement from Fig. 92, shows Eta Carinae as the brightest star at the centre, amongst the white O-stars; the Keyhole is the complex shape to the upper right. Dark dust clouds hide the nebula beyond parts of the Keyhole, but the darkness within the upper circular part of the Keyhole (where the spindle of the key would enter the lock) arises more from a true hollow than from dust. This circle is known as the Carina II Ring and represents an expanding sphere from some unknown energetic event, perhaps stellar winds ejecting material from the surfaces of O-stars within the ring.

The blue tinge to the nebular light between the Keyhole and Eta Carinae is a clue that there are also reflection nebulae in the area. The blue light is in fact a reflection from the dust cloud near the Keyhole of the light of Eta Carinae itself. When Sir John Herschel pictured the Eta Carinae Nebula around 1834–8, he drew this side of the Keyhole much more definitely than it appears today (Fig. 94) indicating that this part of the nebula has faded away. Indeed, this is very understandable, because Eta Carinae itself has faded!

When this southern star was first seen by northern astronomers in 1677, the star was of the fourth magnitude; as can be deduced from its name it was perceived by Bayer as about the seventh brightest star in the constellation. As astronomers became familiar with the star, they realised it was variable, ranging between second and fourth magnitude. When Herschel made the drawing of Fig. 94 it was first magnitude. In 1843 it became the brightest star in the sky after Sirius at magnitude -1, but gradually faded to magnitude 7 or 8. In recent years when the photographs were obtained it has been about

Fig. 92. *(above) Heart of the Eta Carinae Nebula. Within the Eta Carinae Nebula are many clusters of bright stars, some formed relatively recently. Trumpler 14 is the name of the brightest cluster, upper right of centre. Eta Carinae itself is the central bright star, east (left) of a dark cloud which is embedded in the nebula. The Y-shaped broad lanes of dust show no evidence of interaction with the Eta Carinae Nebula – their edges are diffuse, not bright rimmed – and they appear to be foreground dust clouds.*

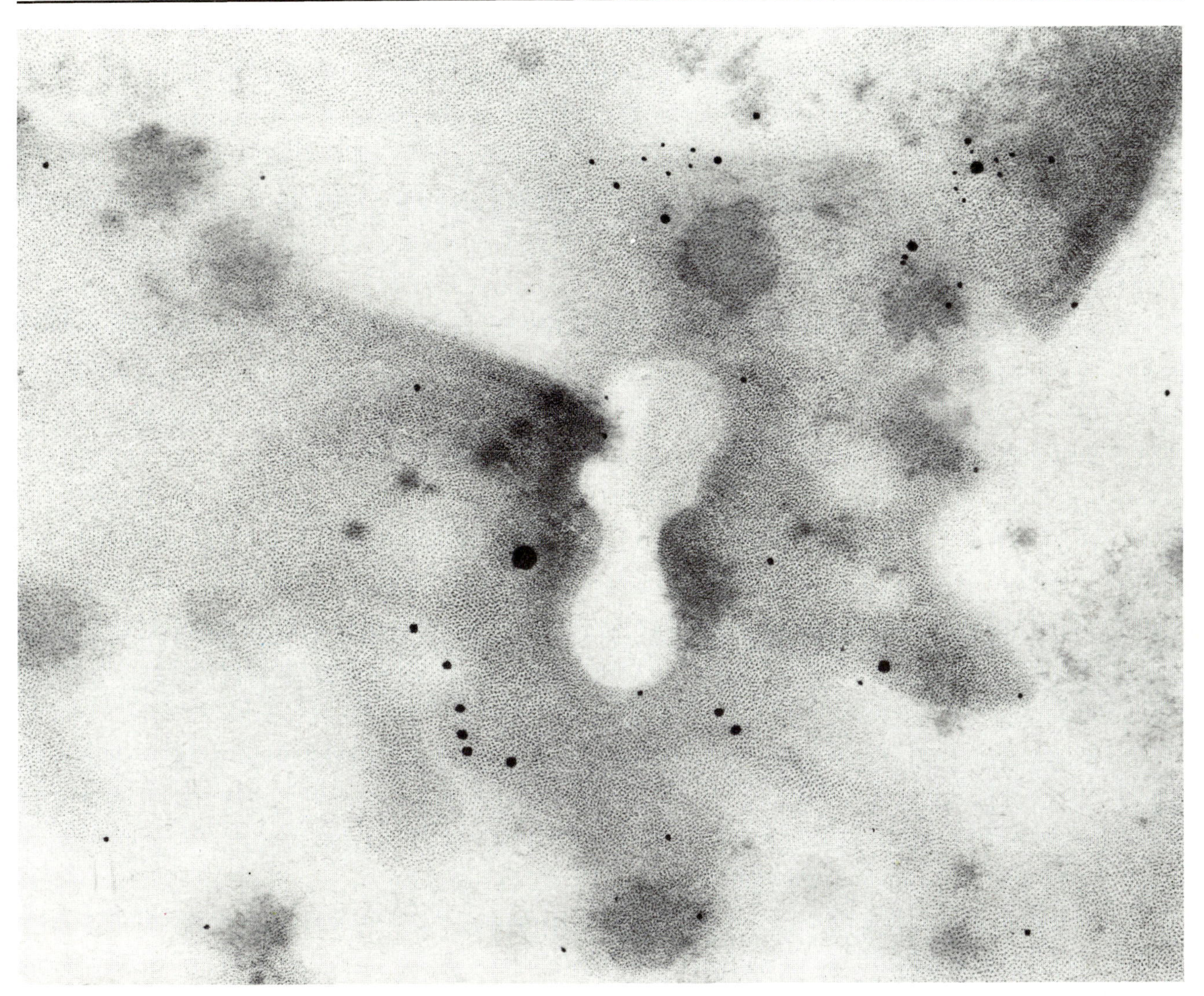

Fig. 93. *(left, below) Homunculus. Surrounding Eta Carinae lies a curious nebula hiding the star within. The small bright orange nebula known from its shape as the Homunculus is seen at the centre of the photograph. Faint wisps of nebula surround the Homunculus. To the right of the Homunculus is the Keyhole. The bright rims encircling the Keyhole are seen well in this enlargement from Fig. 92. The blue colour of the parts of the Keyhole nearest Eta Carinae shows this area to be starlight, reflected off the dust cloud, which reveals its presence to the south in a third way, namely by absorbing light from the stars and nebulae beyond.*

Fig. 94. *(above) The Keyhole in 1840. Clear evidence of change in the Keyhole Nebula is seen by comparing Sir John Herschel's drawing of the Eta Carinae Nebula with a modern photograph (Fig. 93). It is evident that the bright nebula he saw to the west (left) of the Homunculus has largely disappeared. Confidence in the accuracy of Herschel's draftsmanship is engendered by comparing the star positions and details in the nebula with the photograph. The fading of the Keyhole at the same time as Eta Carinae is a further piece of evidence that its light is reflected from that star.*

magnitude 6.2. Thus the star faded by a factor of 100 between the Herschel drawing and Fig. 93, and it is not surprising that the lower part of the Keyhole, the part which is a reflection nebula, has faded to about 1% of its previous brightness.

Even these changes do not, however, represent the most startling facts about this curious object. As can be seen from Fig. 93, the image of the star Eta Carinae is decidedly non-stellar! It is a red-orange oval-shape, a little nebula discovered by Innes 70 years ago and described as having the shape of a little fat man with stubby arms and legs; the little nebula is called the Homunculus (or manikin). The star Eta Carinae itself lies unseen within the nebula, hidden by a close ring of dense dust; its starlight leaks out of the ring and is scattered by dust in the Homunculus, which acts as the principal star in the Keyhole reflection nebula. The Homunculus is expanding at high speed – about 500 km/s. Today the Homunculus is about 12×8 arc seconds, with a rich orange centre and frothy yellow-white surround. Just outside the dense Homunculus is a fainter patchy outer ellipse of nebulosity (just visible in Fig. 93) which was discovered by David Thackeray in 1948 while he was testing the then newly installed Radcliffe 74-inch telescope in South Africa. Each zone of nebulosity around Eta Carinae represents one of a series of stellar explosions occurring in the 1840s, and correlated with the star's spectacular brightening.

The fading of the star since 1843 is more apparent than real. In 1968–9 Gerry Neugebauer and Jim Westphal found that at infrared wavelengths Eta Carinae is the brightest star in the sky. Taking into account the infrared radiation, Eta Carinae is as bright now as in 1843. The optical energy which was visible in 1843 is being soaked up by the dusty nebula formed then and re-emitted as infrared radiation.

In all probability it is the mass of Eta Carinae which is the cause of these curious phenomena. From its energy output we know that Eta Carinae is unusually massive, and stars of mass greater than 100 solar masses cannot form and remain stable. The instabilities in Eta Carinae have caused it to shrug off two or more shells, each an attempt to reduce its mass and become stable. If, as seems likely, it cannot achieve this end, Eta Carinae may explode soon as a supernova, one of the greatest explosions mankind will ever witness.

A gallery of H II regions

The Lagoon Nebula, M8, is an H II region containing the cluster of stars NGC 6530.

The Lagoon Nebula derives its name from a wide and very dark dust lane which appears to wind its way through the nebula. Fig. 95 is a lightly-printed AAT plate which clearly shows the striated dust lane running diagonally across the picture. At the heart of

Fig. 95. *Lagoon Nebula. Listed as Number 8 in Messier's Catalogue, this nebula derives its name from the dust lane which winds its way across the face of the emission region. This short exposure made in red light shows that lane and the dust in discrete globules silhouetted within the bright nebula. The Hourglass Nebula is the brightest region of this print.*

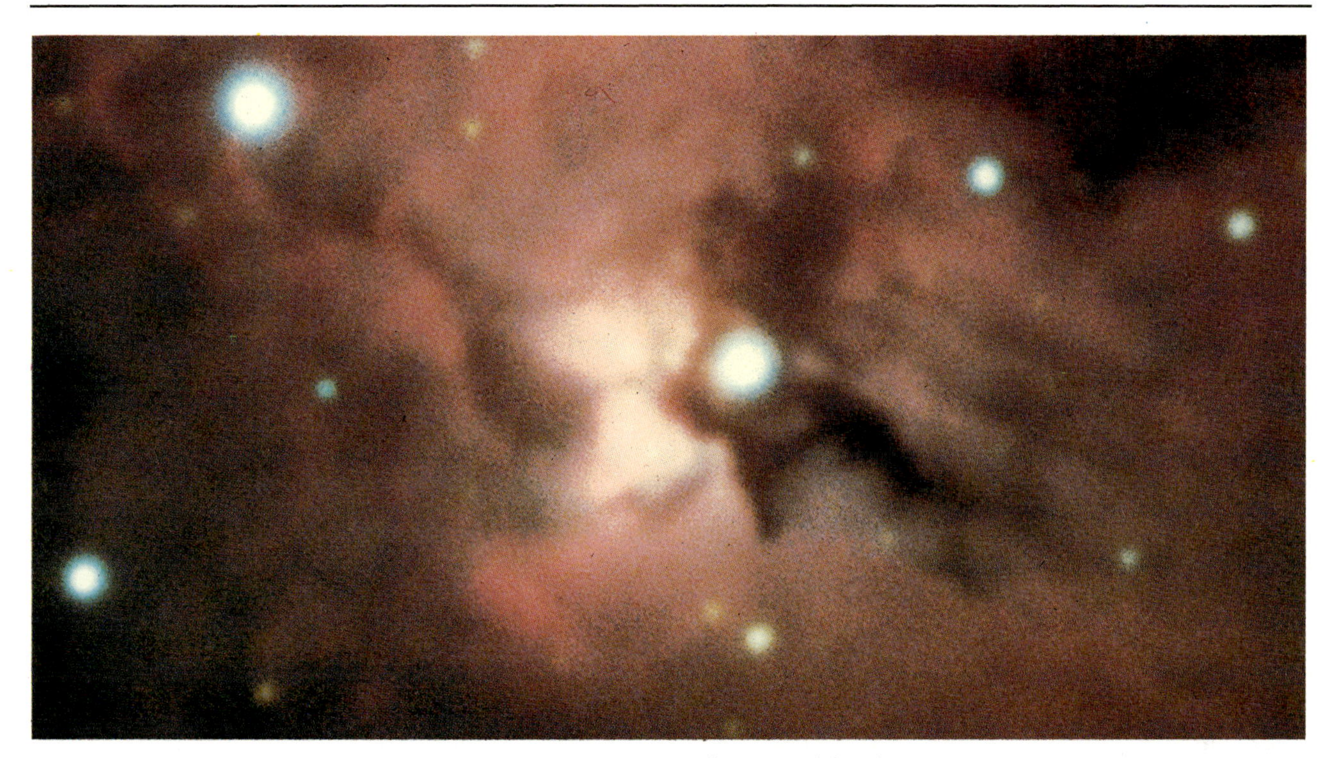

Fig. 96. *Hourglass Nebula. At the heart of the Lagoon Nebula, the brightest part of Fig. 95, lies the tiny Hourglass Nebula, a very bright emission nebula associated with the nearby star named, after its discoverer, Herschel 36. This object is one of the youngest stars known, with an age which has been calculated to be less than 10 000 years.*

the Lagoon Nebula, to the right of the dust lane, is the tiny Hourglass Nebula (Fig. 96). It was discovered by William Herschel's son John. Immediately adjacent and to the right of the Hourglass is a star which Herschel (1840) numbered 36 and it is this star which is exciting the Hourglass. This nebula is extremely young, probably less than 10 000 years old and is about half a light year across.

A wider view of the Lagoon nebula (Fig. 97) shows it to be a complicated H II region with the cluster of stars NGC 6530 lying off centre of the nebula. It, and presumably the nebula, are 4500 l.y. away. Its two brightest stars, 7 and 9 Sagittarii, were the exciting stars of the nebula, but now the brightest part lies west of centre near the Hourglass. The powerhouse of the H II region has shifted to the newer star, which is emitting more ultraviolet light than are the older stars in the cluster.

The colour photograph gives the impression that colder, darker material is pressing around the periphery of the nebula, obscuring the distant Milky Way stars in a zone around the H II region. Globules within the nebula stand prominent. They are often called Bok Globules after the Dutch-American astronomer, Bart Bok (Bok and Reilly, 1947) who first drew attention to them. Driven by the heat and illuminated by the radiation of the young stars which the hydrogen cloud has spawned, the nebula is expanding to surround and engulf outlying concentrations of dust. The sharp-edged rims of some of these dust globules can be clearly seen on the enhanced photograph. Compression of such globules may continue the cycle of star formation and give rise to yet more cluster members. Similar features stand out against the Eagle Nebula surrounding the cluster of stars Messier 16 (Fig. 98). For some reason the nebula, which can be distinctly seen to the south of the cluster, in 6- or 8-inch telescopes, was unremarked by most observers before its discovery by photography, although Messier did refer to a 'faint light' enmeshing the cluster stars.

Another extreme example of compression is shown in Fig. 99, the Cone Nebula. Part of the enormous cloud of dust, gas and stars is NGC 2264 (Fig. 100); the straight sides of the Cone Nebula point directly to a dusty infrared-emitting young star enveloped by the flow of the H II region around it. The curious texture of the nebula in Fig. 100, like walnut grain, and the mixture of red, blue, yellow and brown hues show the chaotic nature of the nebulosity, which consists of H II regions, reflection nebulae (see Chapter 7), and nebulae formed by colliding gas, mixed with young bright stars, which provide the driving energy of the complex. In contrast, the nebulae NGC 6334 and 6357 (Fig. 101(*a*,*b*)) have a colour which is almost completely uniform across the whole of the nebulae and is a dull, deep red. Only NGC 6357 shows a brighter region and this, compared with most other H II regions, is very small and quite yellow in

Fig. 97. *M8 and NGC 6530. This picture of the Lagoon Nebula was made from positive copies of three black and white plates from the UK Schmidt Telescope printed through an unsharp mask to emphasise the dust lanes, globules and bright rims in the nebula, and combined into a colour photograph by addition. NGC 6530 is the designation of the star cluster within the nebula M8; in the NGC Catalogue M8 itself is NGC 6523.*

Fig. 98. *Eagle Nebula. M16 (NGC 6611) is a cluster of young stars which formed about 2 million years ago from the gas and dust which surround them. The dark intrusions visible in the nebula might one day collapse into yet more stars. Bright, red H II regions like the Eagle Nebula surrounding M16 are usually found in the spiral arms of galaxies and are a certain sign of recent and, in this case, continuing star formation.*

Fig. 99. *Cone Nebula. The straight sides of the Cone point into the cluster of stars NGC 2264. In this cluster were discovered for the first time* pre-Main Sequence stars *which were still contracting from the interstellar medium and had not yet arrived in the equilibrium state of the main sequence of the Hertzsprung–Russell diagram. The haloes around the brightest hot stars in the cluster show on this photograph as blue, although the images of the stars themselves are saturated, appearing equally bright throughout the whole spectrum and thus seeming white.*

Fig. 100. *S Monocerotis. The brightest star in the NGC 2264 cluster and visible to the unaided eye, S Mon is at the top of this photo and the tip of the Cone Nebula at the bottom. The brightest part of the nebula is blue, emphasising that it is dust reflecting the light of blue stars nearby, like S Mon, but most colour comes from the recombining hydrogen in the HII region.*

appearance. In addition, neither of these nebulae appears to have the bright blue stars within it necessary to irradiate the hydrogen cloud. It turns out that both of these objects are behind a dust cloud, in the obscuration of the constellation Sagittarius. The cloud is thick enough to affect their colour but not so thick as to hide them completely. Both nebulae are described by Neckel (1978) as heavily reddened. Only the red Hα radiation penetrates the absorbing dust easily. The blue light of the stars and of their reflection nebulae is absorbed, greatly simplifying their appearance.

Stellar bubbles

As the exciting stars of an H II region age, they affect the nebula's shape, often forming it into a hollow bubble. N70 (Fig. 102) is a distinctive circular nebula to the north of the 30 Doradus region of the Large Magellanic Cloud. Its shape shows a symmetry unusual in gaseous nebulae and it was for long regarded as a fossil supernova remnant expanding into a particularly uniform interstellar medium. Uniform expansion into a uniform medium naturally gives a symmetrical simple shape, a sphere. This model seemed convincing for several reasons. For one thing, there was little doubt that the object was a hollow sphere. Such structures have bright rims (because we look through a greater thickness at the edge) and a fairly uniformly bright face. If it was a filled transparent sphere it would be brighter towards the middle, like the Trifid Nebula (Fig. 115) or an elliptical galaxy. Strong supporting evidence for the supernova idea came from the early spectroscopic measurements of the velocities of the filaments in the sphere, and from the same source, the relative abundance of certain of the ionised species within the shell. Radio observation too suggested a supernova origin. A cluster of very hot, young, stars near the centre of the sphere would be just the kind of place to find a massive star of the type likely to explode as a supernova.

Fig. 101. *NGC 6334 (a) (above) and 6357 (b) (top right). The deep red colour of these two nebulae in Scorpius contrasts with the mixed colour of the Cone Nebula or S Monocerotis because of the absorption of blue light.*

Although the supernova idea was for a while accepted, new observations, particularly the spectral index at radio wavelengths (a measure of radio-colour) began to undermine this comfortable assumption. Furthermore, detailed spectroscopic observations showed that the shell was not expanding in exactly the way one might expect if it was the result of a single central explosion. The brightest parts of the N70 shell, those expanding at right angles to our line of sight, would be expected to have little or no velocity in our direction while the centre (i.e. faintest) parts of the bubble should show two distinct components, separated in wavelength as a result of the Doppler shift (p. 177) as the front of the bubble expanded towards us and the back moved away. In fact, only

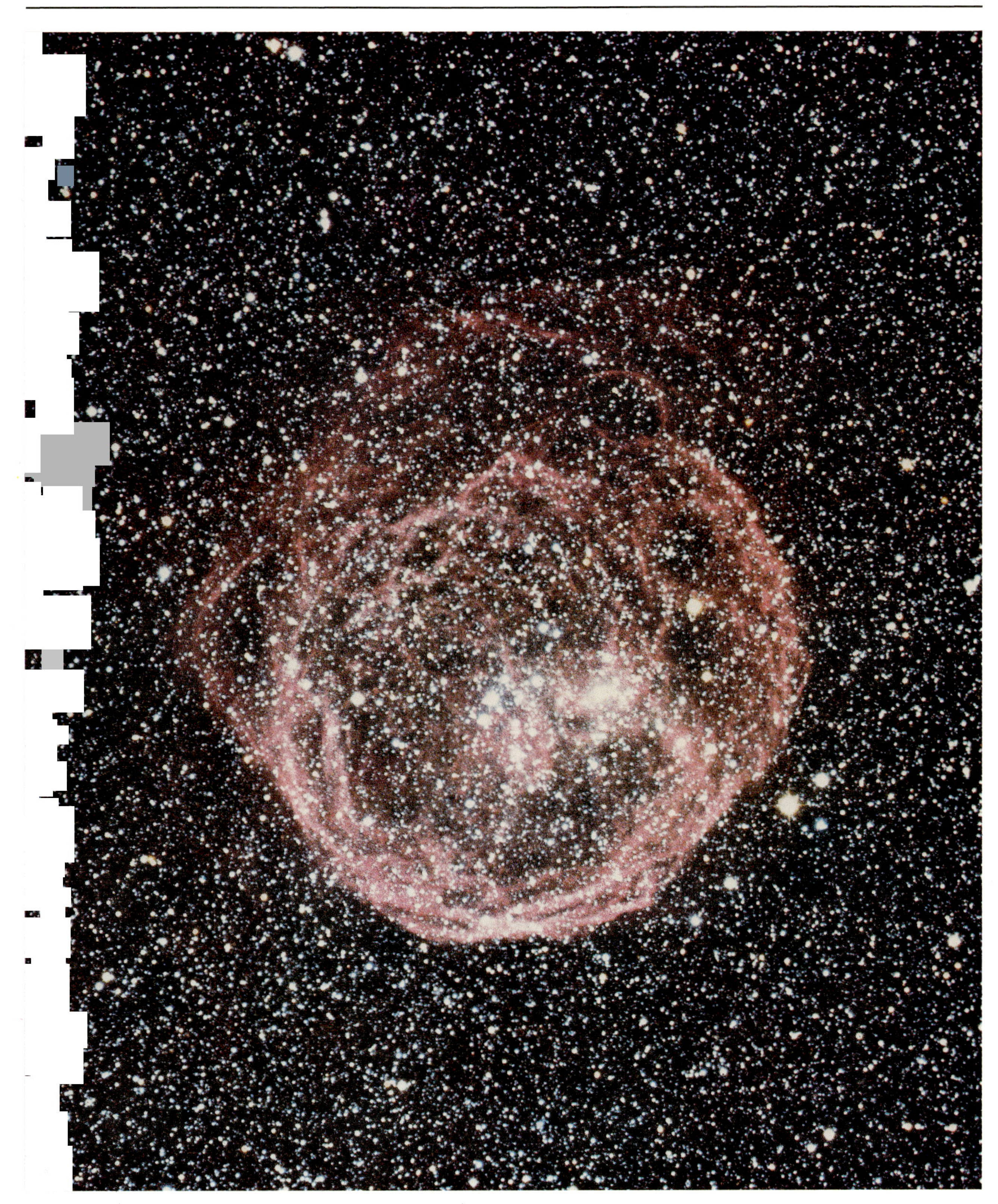

Fig. 102. *N70. A distinctive circular nebula and the seventieth nebula in the catalogue compiled by Henize, this was long thought to be a supernova remnant. This cosmic bubble is now believed to be caused by intense stellar winds from the hot stars visible in its interior. The field stars in this photograph, unlike those in most of the photos in this book, are not in our Milky Way, but in a glaxy outside ours, the Large Magellanic Cloud.*

very small and random velocity shifts are seen. Surprising too was the lack of X-rays from an object surmised to be a supernova remnant. Several X-ray sources are known in the Large Magellanic Cloud (LMC) but N70 was not one of them.

All of this evidence and some new observations made on the Anglo-Australian Telescope have been sifted by Mike Dopita and his colleagues (1981) from Mt Stromlo Observatory in Australia to construct a new model which appears to fit all the observed facts.

They suggest that the power source of the nebula is alive and well, living in or near its centre, and is a star or stars similar to the energetic Wolf–Rayet stars described in Chapter 5. A distinctive feature of these stars is their vigorous stellar wind, if wind is quite the right word for velocities of at least 3800 km/s. According to Dopita *et al.*, the observed almost spherical nebula is a quite stable interface between hydrogen gas originally infalling towards the centre of star formation (the bright off-centre cluster) and the vigorous outflowing stellar wind. The energy involved in this cosmic balancing act can be gauged from the size of the bubble. N70 is 360 l.y. in diameter, about 100 times the distance from our Sun to the nearest star. It is however not the biggest of these impressive bubbles, many of which are found in and around the LMC (see Figs 138–140), but its shape and origins make it one of the more interesting emission nebulae.

Rosette Nebula. NGC 2237–39 is an H II region containing bright stars clustering in its distinctive central cavity. Clearer than the central hole in the Orion Nebula (which we see as M42), this nebula is not so markedly spherical as N70. It comes somewhere between them in age and in the energy with which the central stars are blowing the gas outwards.

Chapter Seven

Starlight and dust

Scattering by dust grains

The most varied colours in astronomy are to be seen in reflection nebulae, Fig. 112. These are clouds of dust found near the central plane of the Galaxy into which, by chance or by generic relationship, shines the light of bright nearby stars. Reflection nebulae are in fact part of the story of interstellar reddening (Chapter 3). Light waves from a star, in passing through a dusty cloud and interacting with a grain of dust, are either absorbed or deflected (for which the technical term is *scattered*). Light which is scattered into our line of sight enables us to view the dust cloud. We call it a reflection nebula.

These same two processes of absorption and scattering occur when sunlight passes through the Earth's atmosphere. When the Sun is high in the sky about 90% of its light passes directly through the air to the ground. About 10% interacts with air molecules, dust particles and water droplets and is absorbed or scattered. The scattered light reaches our eyes from directions away from the line of sight to the Sun and brings with it messages about the atmosphere – clear blue sky, white fine weather clouds, dark thunderclouds, smokey haze, and so on.

Astronomers still argue about the relative proportions of the two processes, absorption and scattering, by which light is lost on its journey through the interstellar medium. Less than half the light lost, probably as little as 20%, is actually absorbed by interstellar grains. (This small proportion however, serves to warm interstellar grains from the near-zero absolute temperature of the dark depths of interstellar space to as much as a few hundred degrees Kelvin.) More than half of the starlight lost in its journey is therefore scattered.

Through thick and thin

Interstellar space is mostly empty, so that the chance of the scattered light wave interacting with another grain is relatively low. This makes the mathematics of the study of scattering conveniently simple. The case where there is only one interaction is called a *single-scattering* process and the dust cloud is said to be *optically thin*. But there are thick dust clouds in the Galaxy and these often make interesting cases to study. The mathematics is often difficult in a *multiple-scattering* process when a dust cloud is *optically thick*. In both cases the degree of scattering depends on the wavelength of the light involved, so that interesting colours arise in the scattering process. We can expect from our ordinary experience that, although the behaviour of the light in the two processes is similar, the appearance of the interstellar clouds might depend on the number of scatterings. The blue of a clear sky at midday is a result of a single-scattering process because only about 10% of sunlight is scattered by the molecules in the 30-km vertical thickness of the Earth's atmosphere (Fig. 103). The chance that a second scattering will occur is again 10%, so only 1% of the sunlight (0.1 × 0.1) is involved in two scatterings. Similarly 0.1% of the sunlight is involved in three scatterings, and so on. The contribution of the multiple scatterings at noon is thus negligible. At sunset however, the Sun's light passes to the Earth's surface slantwise through as much as 400 km of

Fig. 103. *Noon. When the Sun is overhead it shines through about 30 km of Earth's atmosphere and 10% is scattered into the sky. The sky is blue.*

atmosphere (Fig. 104). The chances of multiple scattering thus become greatly increased, and the effect on the colour of a clear sky at sunset is immediately obvious – the sky has lost its deep blue look, and pale shades of oranges, pinks, yellows and white predominate (Fig. 105). When sunlight passes into a very thick, pure scattering medium, the scattered light will be simply sunlight, not coloured at all; thus we would expect the sky inside very thick clouds or mist or dust storms to be the pale yellow or white colour of the Sun, as indeed it is. This example shows that we can expect a range of colours in reflection nebulae, depending on how thick they are, with paler, white colours resulting from denser material in which multiple scattering occurs.

Rayleigh scattering

The nature of the scattering particles, particularly their size, makes a difference to the way that light is scattered. If the scattering particles are very small compared with the wavelength of light (λ), then the scattered intensity of light is inversely proportional to the fourth power of the light's wavelength (λ^{-4}). This remarkable result was derived by Lord Rayleigh in 1871 in response to experiments by the chemist John Tyndall. He had observed the blue colour of scattered light and challenged Rayleigh to explain it on the basis of his new theory of radiation. The scattering process is known in fact as *Rayleigh scattering* in this case. The inverse fourth-power law means that ultraviolet light of a wavelength 0.35 μm is 16 times more likely to be scattered than red light of 0.70 μm wavelength (Fig. 106). Thus pale yellow light from the Sun which passes into a very clean atmosphere and is scattered by

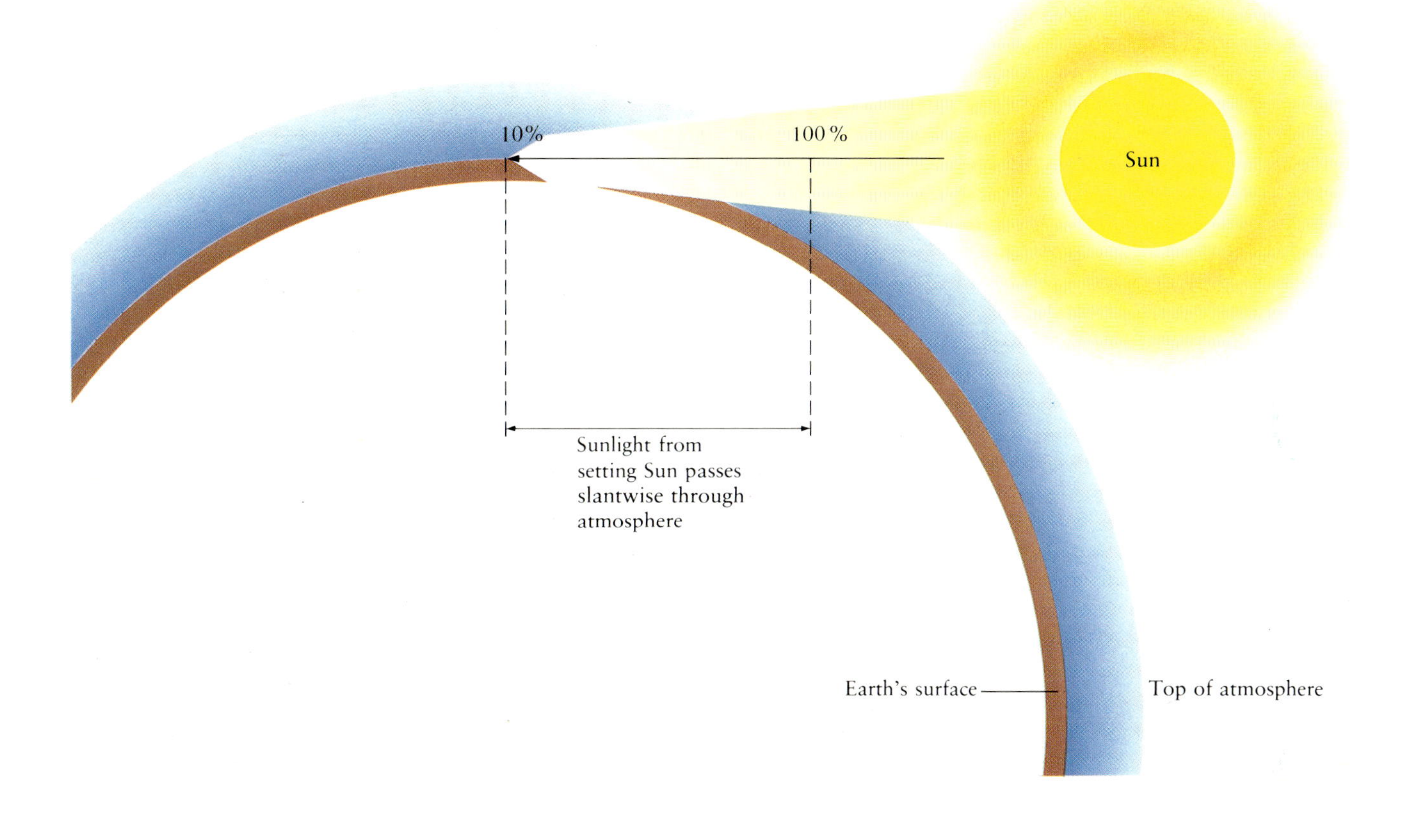

Fig. 104. *Sunset. The path of sunlight through the Earth's atmosphere is much longer when the Sun is setting. More light is scattered, and the sky is whiter.*

very small particles (air molecules, of order 1–10 Å in size) yields a scattered component which is very blue – sky-blue in fact. Because the colour of the scattered light depends on such a steep power law, the colour of the light does not change much if the spectrum of the input light is changed. In fact the Rayleigh-scattering atmosphere of an Earth-like planet orbiting virtually any star will be a more or less intense shade of blue, except that the sky of a terrestrial planet orbiting the reddest stars (cooler than about 2500K, the equivalent of M5 spectral type or so) would be white, or perhaps the palest pink colour (Fig. 107).

If Rayleigh scattering was the scattering process in the interstellar medium all reflection nebulae would be very blue. But are interstellar dust grains small enough? The variety of colours of the nebulae in Fig. 112, particularly the red tint to the nebula around Antares, suggests that they are not.

Scattering by very large particles

If the scattering particles in the interstellar medium were very large and absorbed no light at all, then they would scatter light like little mirrors. Sunlight striking a collection of randomly oriented mirrors would scatter in all directions, like a spot light on a rotating ball of mirrors suspended over a dance floor. The mirrors do not colour the light at all (by hypothesis *all* the light is scattered, none absorbed). Thus the scattered sunlight would be white. Water drops fulfil the definition of 'large compared with the wavelength of light' (drops of rain may be measured in

Fig. 105. *Sunset in a clear sky. As the Sun sets its light shines slantwise through a larger amount of atmosphere and its colour becomes orange, together with the sunlight which illuminates the sky nearby. The sky generally whitens. In this photograph of the Sun setting over the twin-peaked island of La Palma in the Canary Islands, orange sunlight is also reflected from clouds which cover the sea.*

millimetres, drops of mist in tenths of millimetres, fine drops in clouds perhaps in hundredths of millimetres) and they are virtually colourless. Thus clouds of water droplets are white, as we know from experience.

The sky on Mars is coloured pink (Fig. 108) by large-particle scattering from dust grains of limonite. Limonite is a brown oxide of iron. It covers the Martian desert and gives Mars its well-known red colour. Grains of limonite are blown into the Martian atmosphere by sandstorms. The actual air of Mars is very thin and scatters almost no sunlight, so there is no blue sky. Indeed if there were no dust, the stars would be perpetually visible in the Martian sky even though the Sun was above the horizon. However, sunlight is scattered and absorbed by the large limonite particles. Since limonite is not colourless, indeed limonite grains preferentially absorb blue light and act like mirrors tinted red, the Martian sky is pink.

If large-particle scattering were the scattering process in the interstellar medium, reflection nebulae would have the exact tints of the stars whose light they scattered or the colour of the grains themselves if they absorbed, as well as scattered, light.

Scattering in the interstellar medium

The particles in the interstellar medium are neither very small compared with the wavelength of light, nor very large. They are in size not much different from the wavelength of light. A typical grain is about 1 μm in size. Thus they give rise to a form of scattering that in some senses is intermediate between Rayleigh scattering and large-particle scattering. This is known as Mie scattering after Gustav Mie who, in 1908, laid the groundwork for its study. However, the precise details about the dust grains become very important in Mie scattering, whether they are spherical or elongated, what they are made of, etc., and the exact calculations of Mie scattering often defeat the most talented mathematicians with access to the most powerful computers.

We know from the phenomenon of interstellar reddening that starlight which traverses an interstellar cloud is reddened, so that more blue light than red light is removed from the starlight. If this light is scattered to us then the light will be generally blue, so that the amount of scattered light is inversely proportional to some power, α, of the wavelength, λ. For Rayleigh scattering, $\alpha = 4$; for large mirrors, $\alpha = 0$; we expect α to lie between 0 and 4. Observations show in fact that α is approximately unity, $\alpha \approx 1$. The difference between the scattering laws is large, as shown in Fig. 106, but we can still expect reflection nebulae to be blue like the blue sky. Because the λ^{-1} law is not so steep as Rayleigh scattering we would not expect it to be so universally true that all reflection nebulae are blue no matter what star shines into them, unlike the earlier statement that nearly all terrestrial planets have a blue or white sky no matter what star they orbit. Fig. 109 shows that we would see as wide a variety of colours as we do for stars. In every optically thin case the reflection nebula would be somewhat bluer than the illuminating star, as can be seen by a close comparison of Fig. 109 with Fig. 107(*a*) – the curves in Fig. 109 are tilted up to the blue more than those in Fig. 107(*a*).

Fig. 106. *Scattering laws. Suppose that white light falls into three scattering media. The 'neutral' scatterer has no effect on the colour of the scattered light – it still has a spectrum given by the flat curve labelled λ^0. A Rayleigh scatter, on the other hand, scatters the blue light much more and the red light much less – the scattered light has a spectrum labelled λ^{-4}. Interstellar particles are between these extremes and the light they scatter has a λ^{-1} spectrum. (The vertical axis shows the scattering function measured in intensity units relative to the intensity at 5500 Å.)*

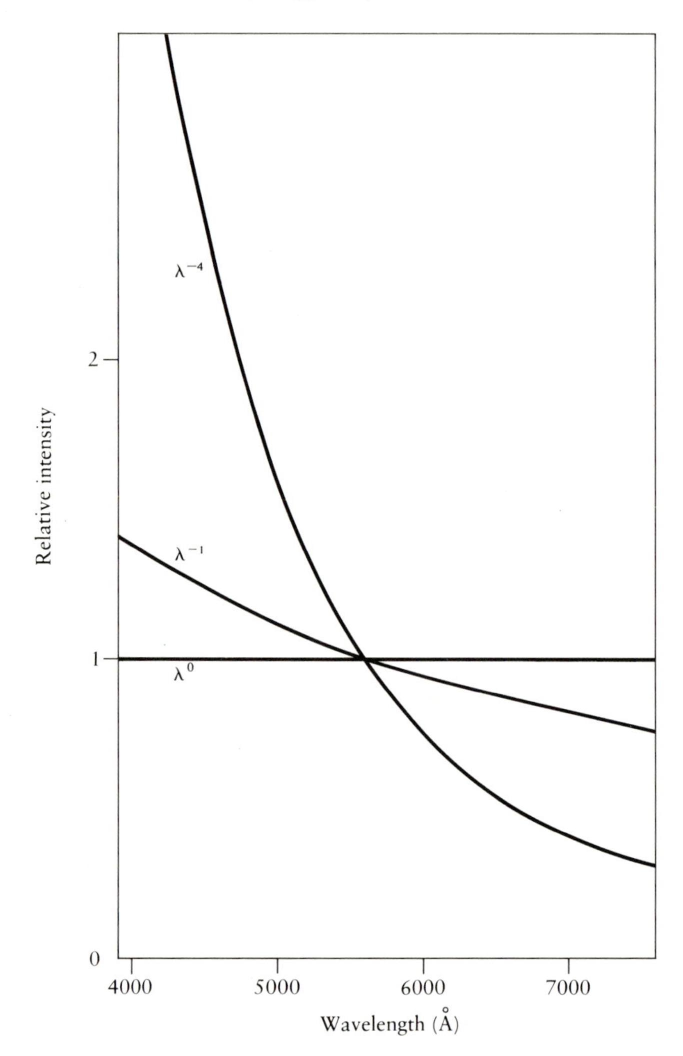

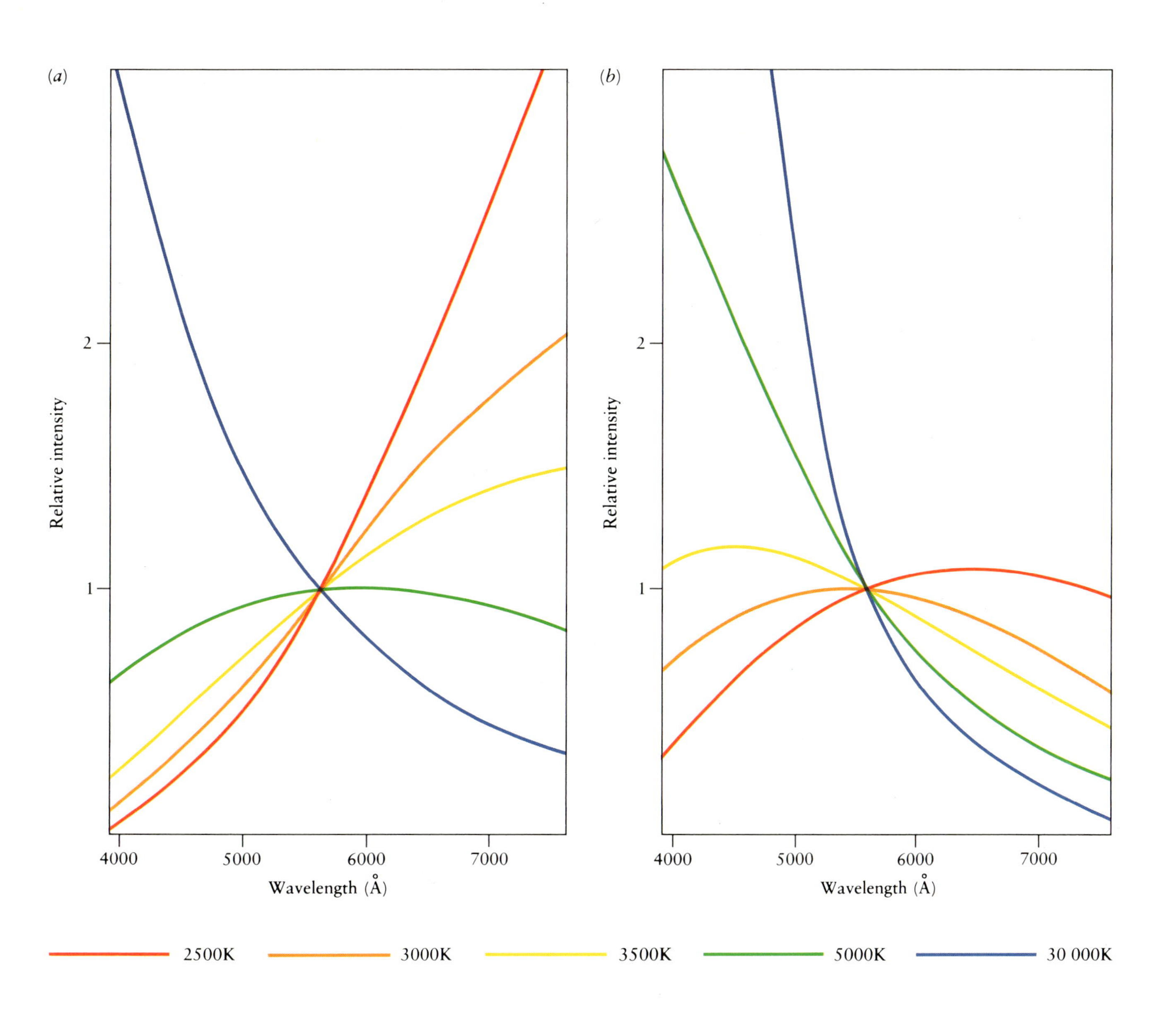

Fig. 107. *Blue skies. If the light input into a Rayleigh-scattering medium is not white, but has a black-body spectrum, the scattered light is always made bluer by the scattering.* (a) *(left) Spectra of black bodies at different temperatures.* (b) *(right) Spectra of scattered light from these black bodies after scattering by a* λ^{-4} *medium, in other words the spectra of the skies of Earth-like planets with different suns. Note how nearly all the curves in* (b) *rise to the blue – the skies of Earth-like planets are basically blue, no matter what the temperature of their sun.*

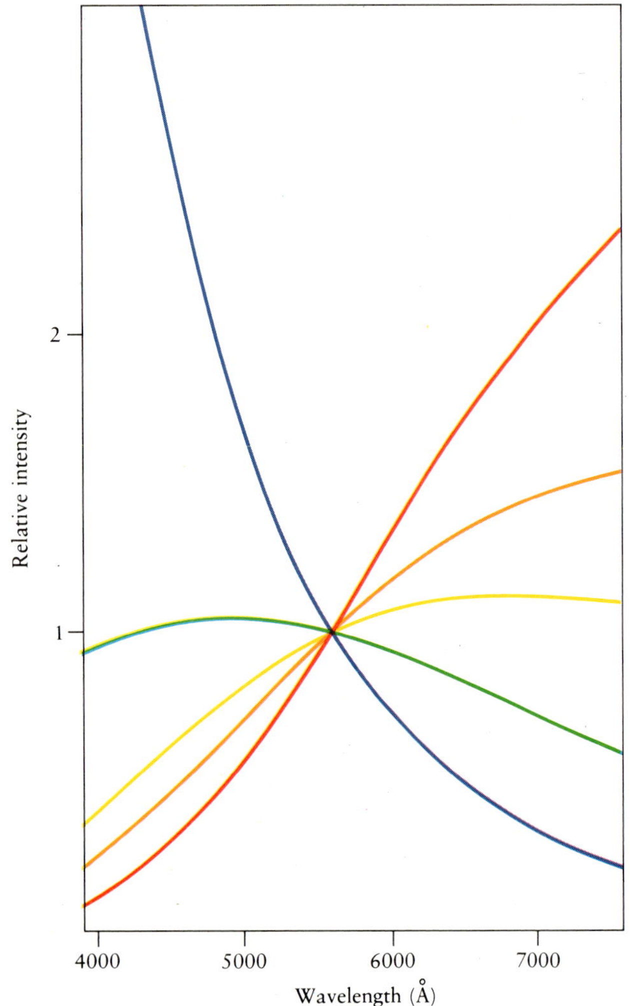

Fig. 108. *(above) Pink sky of Mars. The Viking Lander photographs of Mars show a rust-coloured sky due to sunlight reflected by limonite dust. The pictures were transmitted from Mars in three-colour separated form. When added in the photo-lab the first trials were balanced so as to give a blue sky like Earth and released this way to the Press. Correct rebalancing revealed the unexpected pink colour.*

Fig. 109. *(left) Colours of reflection nebulae. If the black-body spectra of Fig. 107(a) are scattered by the interstellar medium (λ^{-1} law), the spectra of the scattered light become like the ones in this figure (colour coding represents original black-body temperatures; see key in Fig. 107(a)). These curves slope up to the red or up to the blue, and represent a variety of colours.*

In fact most reflection nebulae are formed by the brightest stars shining into nearby dust. The brightest stars are usually blue giants (see the HR diagram of the Pleiades, Fig. 66). Moreover, dusty clouds are usually found in the plane of the galaxy near to such massive blue stars. Thus most reflection nebulae are blue, although they need not be.

It is because of the chance coincidence that grains in the interstellar medium are the size of the wavelength of light that the colours of reflection nebulae are as varied as they are.

Forward and backward scattering

Effects of optical depth (i.e. multiple scattering) may show up in pictures of reflection nebulae in that as light penetrates deeper into a nebula before being scattered in our direction, it will become whiter; of course some blue light will have been removed from the starlight, so starlight scattered from the outer parts of a reflection nebula can even be red. We would see a trend across the nebula from the middle to the edge such that the colour became whiter further away from the central star (Fig. 110). In most scattering processes by small particles, the scattering is 'forward-peaked' – that is to say the light from the star is deflected only

Fig. 110. *The Sun shines through dense smoke over Sydney from bushfires in the Blue Mountains. The Sun itself is reddened by the smoke. Sunlight scattered from the orange sun is deeper orange in the vicinity of the Sun where only one scattering has re-directed sunlight to our eyes. Further from the Sun, the colour pales because we see a re-mixture of light after multiple scatterings.*

slightly from its direct path. Thus mist is most visible in the direction toward the Sun and not so dazzling in the direction away from the Sun. Similarly (Fig. 111) raindrops on a window blur the outline of what is beyond. The effect of forward scattering in most reflection nebulae is to give them the appearance of a halo symmetrically disposed around the star. Of course if the star is on the near side of a cloud of interstellar grains, the reflection nebula can only be seen by backscattering, in which the starlight is turned through 180°. Backscattering is so different from forward scattering that it adds interesting effects to the colours seen (Roark & Greenberg 1967), although it seems that no particular nebulae have been identified with these possible geometrical differences.

Most reflection nebulae are more akin to the halo seen around the lights of a car as it approaches through fog, than to the backdrop of a theatre's stage illuminated by footlights.

Fig. 111. *Raindrops on a window. Lying below the line of sight to the top of the wall beyond, the drops scatter skylight up into the camera and show as white specks against the grass, while those lying above the top of the wall image the darker grass and show as dark specks against the sky. There is a narrow band within which the raindrops show, centred on the line of the wall. The band represents the deflection of light by the raindrops, which scatter it forwards into the camera.*

TABLE 12. REFLECTION NEBULAE ON THE SCORPIO–OPHIUCHUS BORDER

Star			Nebula		
Name	Spectral type	$B-V$	Name	Colour coded by van den Bergh	Colour index
σ Sco	B1	0.14	vdB 104	Intermediate	−0.16
ρ Oph	B3	0.24	IC 4604	Very blue	−0.15
−24° 12684	B2	0.85	IC 4603	Moderately blue	−0.33
22 Sco	B2	−0.12	IC 4605	Very blue	−0.36
Antares	M1	1.83	IC 4606	Very red	+0.40

(References: Struve, Elbey and Roach, 1936; van den Bergh, 1966; Racine, 1968.)

The Ophiuchus cloud

Many of these effects can be seen in Fig. 112(*a*), a three-colour additive print from plates taken with the UK Schmidt Telescope. These reflection nebulae are on the border of the constellations Scorpio and Ophiuchus (compare Fig. 112(*a*) with the wider-angle view in Fig. 24). Antares is the bright star in the lower left corner, and the globular cluster M4 is at the bottom of the print. At the top is the Rho Ophiuchi reflection nebula. The area is very dusty, as can be inferred from the many places in the picture where no background stars can be seen. In fact it was here that W. Herschel remarked on the hole in the heavens (Chapter 3) at the centre of the photograph.

The fact that there are such thick nebulae in the area shows that we can expect effects of multiple scattering from optically thick nebulae and the varied colours which this will entail. A key to the nebulae in the picture is given in Fig. 112(*b*).

First of all, observe that to the right of Sigma Sco is a semicircle of red nebula which is the only non-reflection nebula contributing to the coloured image. The red is from Hα emission, the nebula being an H II region excited by Sigma Sco which is the only star in the area which is hot enough to emit sufficient ultraviolet light to ionise hydrogen (its spectral type is B1). To the left of Sigma Sco is a faint bluish haze which is a reflection nebula scattering light from Sigma. The three reflection nebulae in the top half of the picture are also blue. The stars whose light they reflect, while not as hot as Sigma Sco, are also quite blue (Table 12). The colour differences evident on the photograph have not yet been confirmed by measurements. The reported measurements by Struve, Elbey and Roach (1936) and the colours categorised by van den Bergh (1966) of these faint nebulae are too inaccurate to differentiate between the grades of blue seen here. They do confirm however the difference in colour (Table 12) between the blue nebulae and the pink nebula, well seen below Antares. Antares is a red star, as is evident from inspection by the unaided eye, from its name (meaning 'rival to Mars'), from its spectral type, M1, and from its colour index ($B-V=1.84$). The nebula is, however, pink, not deep red like Antares. (The colour of the star does not show on the photograph, as Antares is so bright that its image saturated the three photographic emulsions appearing equal on each and therefore white.) The softening of the red colour of Antares to the pink of the reflection nebula is consistent with the calculations of Fig. 109 which show how the red end of the 3000K black-body spectrum is reduced when scattered by a λ^{-1} medium.

Finally, we should discuss the green, yellow and brown tints that lie between the 22 Sco Nebula and Antares. Since, however, this photograph constitutes new research material it is not clear what the explanation is for these tints. Is the yellow nebula symmetric about 22 Sco? If so it may represent optical-depth effects which would colour the 22 Sco Nebula red towards its extreme edge. Is the yellow nebula part of the red Antares Nebula, again showing some effect which blues the nebula to the edge? Is the yellow nebula a mixture from the blue reflection nebula of 22 Sco and the red reflection nebula of Antares? The green regions must be caused by a spectrum which is strong in the green and weak in red and blue, and are hard to explain. Finally the brown colours are an enigma. Brown is a so-called non-spectral colour. This means it cannot be made simply by combining red, green and blue lights in an additive process – grey, silver and gold are non-spectral colours in the same way. *Brown* represents a psychological response to the texture of a surface, and perhaps in these photographs our minds perceive the striations on the dust clouds as they do the wood-grain on the surface of a violin, and call it brown. Compare for instance Fig. 100, the NGC 2264 nebula.

The colours of Fig. 112 represent an interesting research project.

Reflection nebulae and emission nebulae

The Scorpio–Ophiuchus cloud is remarkable because it contains so many reflection nebulae and so few emission nebulae. The reverse is usually the case and the emission nebulae usually dominate pictures of the Milky Way. This is the case for the remarkable region of Sagittarius seen in Fig. 113 and not previously photographed in colour.

The two main blue objects, NGC 6589 (top) and NGC 6590 (lower) are the reflection nebulae surrounding stars of spectral type B5 and B6, which are not hot enough to photo-ionise sufficient amounts of the hydrogen undoubtedly present. The photograph, made from three AAT plates, contains at least five other distinct but smaller reflection nebulae. The dominant feature is, however, the large patch of photo-ionised hydrogen gas, IC 1283–4, which contributes the characteristic red Hα light to a substantial part of the field. Mixed with the hydrogen gas is an extensive but rather tenuous dust cloud sufficiently reflective to dilute the intense red of Hα emission with a soft blue haze of a superimposed reflection nebula.

Making colour pictures of new areas of sky like NGC 6589–90 is always exciting, mainly because one cannot visualise the end result. Planning the taking of the pictures, we could see that the region was strongly coloured, simply by comparing the red- and blue-light prints of the Palomar Schmidt survey. However, we were not aware of the subtleties or detail in the dusty IC 1283–4 complex or of the extent and structure of the reflection nebulae which dominate the right of the picture. Quite invisible on the Palomar prints was a minute wispy object, just a little to the left of NGC 6589, the lower of the two blue reflection nebulae in Fig. 114. Exactly what this is we do not know, although we suspect it is a Herbig–Haro object, the first visible signs of a proto-star, usually found in dusty clouds. This peculiar loop, just a few arc seconds long and very faint, must await investigation on a spectrograph attached to a large telescope before anything further can be said about it. In the meantime, our illustration, Fig. 114, is about the best finding chart any astronomer could wish for – and probably the first in colour!

Any photograph of the dusty constellation of Sagittarius (Fig. 115) reveals the Trifid Nebula, Messier 20, an object sufficiently complex and

Fig. 112. *Rho Ophiuchi. (a) (left) The glorious colours of reflection nebulae predominate in this photograph, keyed in (b) (above) Only the pink semi circle at lower right around Sigma Scorpii is an H II region. M4 is a globular cluster of stars (bottom centre). The Rho Ophiuchi dark cloud is left of centre, and left of Rho Ophiuchi itself.*

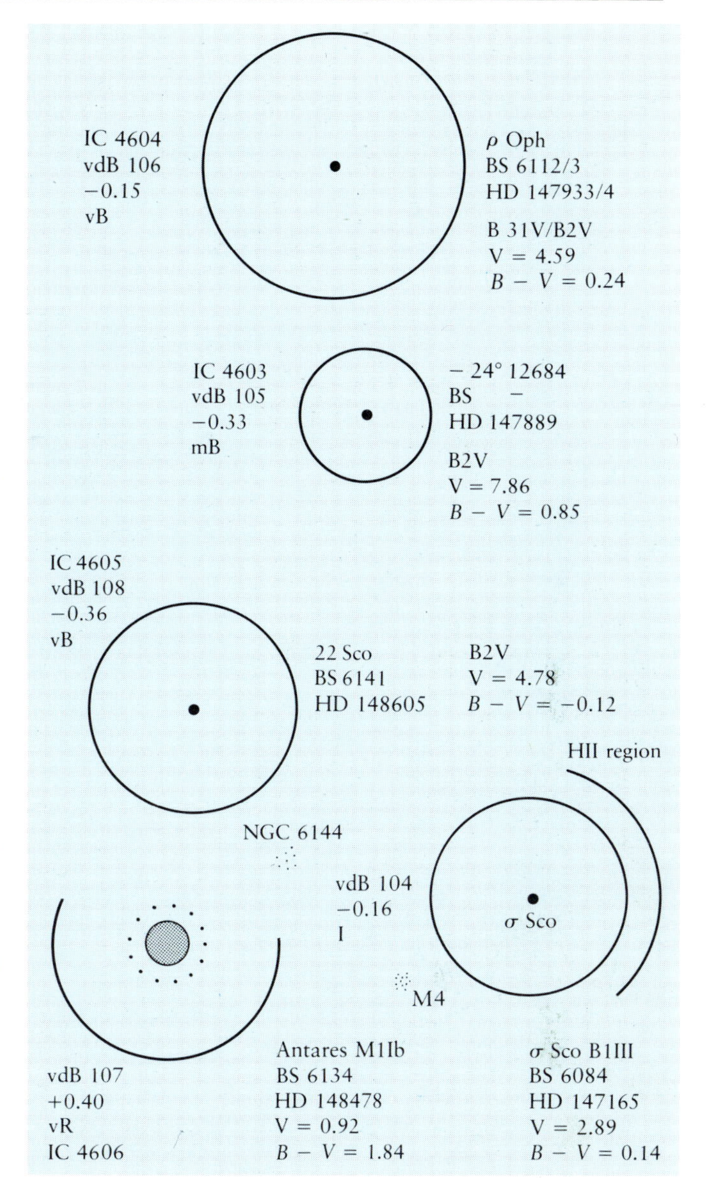

Fig. 112(b). *The nebulae and prominent stars in Fig 112(a) are named in the map above.*

Fig. 113. *(right)* NGC 6589/90. *This colourful region of Sagittarius is known only by the rather dull catalogue names of NGC 6589 and 6590, the lower and upper blue reflection nebulae, and IC 1283–4, the large red H II region. The red of the H II region is here diluted by blue reflection nebulosity from the dust within it, giving rise to the overall characteristic pink colour of nebulae seen in spiral arms of galaxies.*

Fig. 114. *(above) Unidentified object in Sagittarius. Just to the east (left) of NGC 6590, the lower of the two large reflection nebula in Fig. 113, is a tiny worm-like nebula, shown enlarged in this illustration. Its nature is uncertain but it has the appearance of a Herbig–Haro object, often seen where stars are forming. Examination of this curious feature with a spectrograph attached to a large telescope will be required to discover more.*

Fig. 115. *(right) The Trifid Nebula combines red emission and blue reflection nebulosity to produce one of the finest spectacles in the Milky Way. As in IC 1283–4 (Fig. 113) the red Hα emission around the bright central stars is diluted with blue scattered light. These combine to give the deep blue-red of the emission nebula. To the north, dust around a bright but cool star gives rise to a reflection nebula. The nebula derives its name from three radial dust lanes.*

Fig. 116. *(above) Heart of the Trifid. In the centre of the Trifid Nebula is HN70, a small cluster of extremely hot stars which emit sufficient short-wavelength radiation to ionise the whole of the red cloud seen in Fig. 115. It is clear from the bright rim of the dust lane nearest the star cluster that this dust is indeed embedded in the nebula and not merely a line of sight coincidence. The blue cast to the central nebula is reflected light from HN70 which evidently passes behind the absorption lane to the nebula in upper right. Thus, although the dust is embedded in the nebula it has the form of a rope around which light can pass, rather than that of an impenetrable curtain.*

Fig. 117. *(right) Ionisation bounded nebula. The presence of reflection nebulosity completely surrounding the Trifid Nebula was discovered on this three-colour photograph. The gas mixed with the dust thus revealed is not ionised because the ultraviolet light from HN70 is all used up in maintaining the ionisation of the hydrogen in the Trifid Nebula. As soon as a proton and an electron recombine to make an atom, they are split apart by absorbing ultraviolet light from HN70. Such a nebula is called ionisation bounded, and no ultraviolet light from the stars within escapes from it. The faint blue nebulosity, dimly seen in Fig. 115, is intensified by unsharp masking.*

beautiful to warrant several pictures where normally one might suffice. Fig. 115 was taken on the AAT on colour negative film. At the centre of the almost spherical red cloud is what appears to be a pair of bright stars. A closer look (Fig. 116) reveals a cluster of four stars, two of which are spectroscopic binaries. The hottest star of the six is HD 164492, of spectral type O8, which implies a surface temperature of more than 30 000K. The ultraviolet light emitted by this star is sufficient to photo-ionise a huge cloud of hydrogen gas. The red light of Hα is emitted as the ionised atoms recombine into neutral hydrogen. To the north of the ionised gas is a bright but much cooler star, HD 164514 (spectral type F5), with a surface temperature of around 6500K. Its light is scattered by the surrounding dust particles to give the beautiful blue reflection nebula which makes this object particularly impressive.

In reality the whole area is a mass of gas and dust, and three-colour addition of unsharp masked photographs from the UK Schmidt Telescope (Fig. 117) reveals that the area of reflection nebulosity

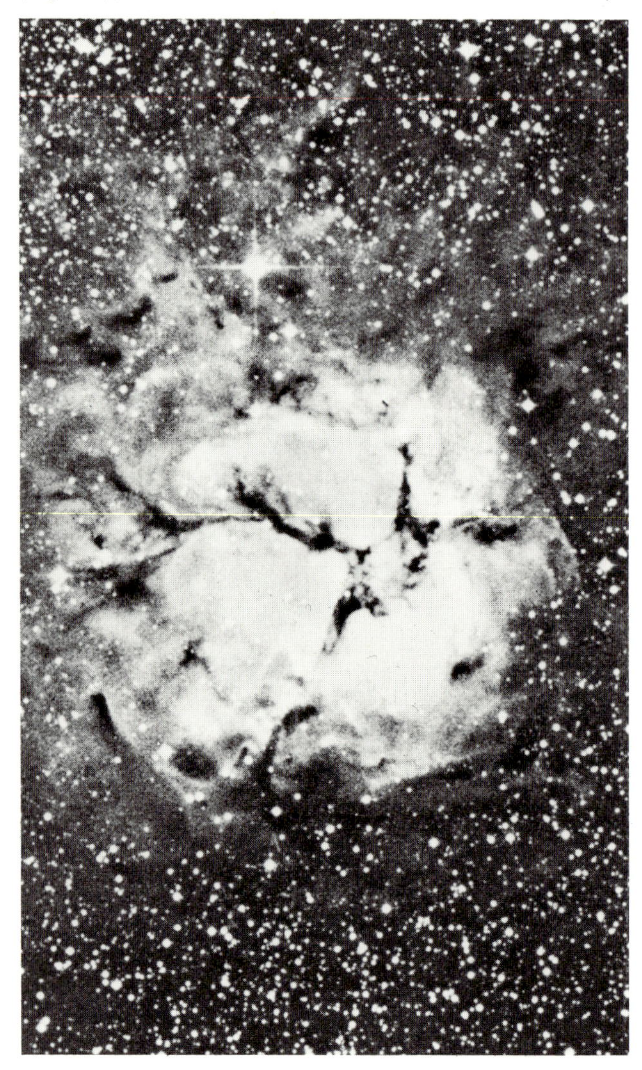

Fig. 118. *(left and above) Trifid reflection nebula. High contrast printing techniques applied to very deep plates from the UK Schmidt prior to their combination into the colour picture of Fig. 117 underline the relation between the reflection and emission regions. The blue-light plate (above) reveals that the reflection nebula extends throughout and well beyond the emission region and has no well-defined boundary, whereas the Hα emission (left) terminates quite abruptly.*

Fig. 119. *(right) IC 1274–5. About 2° to the east of the Trifid Nebula and probably part of the same dusty cloud is this beautiful band of nebulosity set against the yellow stars of the Galactic bulge. The nebulosity in the upper part of the picture (IC 1274–5) is clearly part of a larger obscuring cloud (Barnard 91) of variable density. In the southern part of the picture the bright reflection nebula is NGC 6559 which seems to be associated with a sinuous dust lane bisecting the HII region.*

Fig. 120. *Toby Jug Nebula. The nebulosity around this star, which is known only by its catalogue designation HD 65750, is due to light reflected by dust. The dust is thought to be mainly particles of silicates (sand!) condensed from material which the star is losing. The stellar nature of the central object is revealed by the four diffraction spikes extending from it and circular ring surrounding it. These are artefacts caused by structures within the telescope and halation (internal reflection) from the red-sensitive plate.*

is in fact extremely large and completely surrounds the red ionised gas. The dust lanes which give the Trifid nebula its name originate from the surrounding haze.

The colour photograph demonstrates that there is plenty of dust outside the Trifid Nebula, and in the interestellar medium dust and hydrogen gas go together. Although the Trifid is thus surrounded by hydrogen, its edge is sharply limited. Moreover the nebula is quite symmetrical and looks spherical, despite the clear indications that it is off-centre from the cloud of material in which it is embedded. Why is this? What sets the boundary? The answer lies in the limited power of HD 164492, the star which excites the nebula. The star emits, equally in all directions, ultraviolet photons capable of ionising hydrogen atoms. Each photon travels outwards until it encounters a neutral hydrogen atom which it promptly ionises. The electron thus ejected wanders about in the nebula before finding a proton with which it can recombine to make a neutral hydrogen atom again; but the next ultraviolet photon passing nearby to this atom will promptly ionise it again. Thus there is set up a balance between the emission of ultraviolet photons and the recombination rate of hydrogen in the nebula. The sphere within which hydrogen is ionised is called a Strömgren sphere, after Bengt Strömgren who developed the theory of these nebulae, and the size of the sphere is set by the star's output of ultraviolet photons. If, for instance, the star doubled its ultraviolet output, it could ionise twice as much hydrogen and push the boundary of the Trifid 26% further out in all directions (because $2 = 1.26^3$). Of course, this could only be done if there was neutral hydrogen outside the present boundary; if not, any increase in power would be unused and ultraviolet radiation would escape from the nebula into space.

A nebula outside which there is more hydrogen is called *ionisation bounded*, because the boundary of the nebula is defined by the furthest point at which hydrogen is ionised. The contrary case is called a *density bounded* nebula because the boundary is set by the fall of the density of interstellar material to zero. An example of a density bounded nebula is the Helix (Fig. 75). The colour picture shows that the Trifid Nebula is ionisation bounded, because more unionised material lies outside the boundary.

This point is underlined by comparing the black and white pictures in Fig. 118. These are the blue (right) and red (left) colour separations, which, together with a green-light image (not shown) were combined to make Fig. 117. Notice that the boundary of the red image is sharp, well defined and roughly circular, while the blue reflection nebulosity gradually merges with starry background. It is clear from this comparison that blue reflection nebulosity occurs throughout the nebula but is so relatively weak compared with Hα that it is seen only beyond the boundary of the Strömgren sphere.

Quite nearby in the sky and probably part of the Trifid complex is a much fainter H II region known only by its catalogue designations NGC 6559 and IC 1274–5 (Fig. 119). The full extent of the dust cloud associated with the object is indicated by obscuration of the distant yellow background stars; only a small part appears as a reflection nebula, in the upper part of the photograph. Like M20, this reflection nebulosity surrounds an H II region, IC 1274. The dark cloud to the east (left) of IC 1274 is known as Barnard 91.

To the south, the larger H II region NGC 6559 is crossed by a sinuous dust lane.

Toby Jug Nebula

Finally, in these series of reflection nebulae we include one in which it is clear from the relatively symmetrical shape that the star created the nebula. In the Rho Ophiuchi and Antares nebulae, as in most reflection nebulae, it seems likely that the stars and nebulae find themselves together in space by chance, the star having wandered from its birthplace and found a random nebula to illuminate. HD 164492, on the other hand, was probably formed from the interstellar cloud which it is now ionising to give the Trifid Nebula. In IC 2220 (Fig. 120) the nebula is actually formed by the star. An observer of this nebula might be struck by its similarity in shape to a Toby Jug.

IC 2220 is distinctive not only for its shape. Almost hidden at its centre, but visible on this photograph because of the four diffraction spikes which project from it and the ring which encircles it, is the illuminating star (the ring represents a reflection of starlight in the photographic plate and is not really in the sky). This star is known by its unglamorous catalogue designation HD 65750.

The nebula is unusually bright to be reflecting HD 65750 alone, but there is no other source of illumination. This suggests that the dust which gives rise to the reflection may be of a different kind from that in most nebulae. This possibility is made more credible by the unusual nature of HD 65750. The central stars of reflection nebulae are usually young, hot and blue; HD 65750 is old, cool and red, and is losing gas from its outer layers in a steady trickle. Dust grains rich in silicate minerals are condensing from the gas as it cools, and it seems possible that most or all of the dust and gas of the nebula have been produced from material lost by the star. The process is a form of mass loss as in the unstable stars in Chapter 5. Because the dust grains are recently formed and from just one star, they are probably not typical of grains in the interstellar medium, which are mixed up from lots of stars and have had longer to grow to a larger size.

The yellow colour of the nebula is a consequence of the red light from the star being scattered by the small grains.

Chapter Eight

Colours of galaxies

Galaxies and the Universe

A galaxy is a collection of stars and gas, typically tens of thousands of light years in size, and usually separated from its nearest neighbour by millions of light years. The separation of galaxies is so large that in photographs of the sky most of the photograph is black – the galaxies are spots of light quite distinct from each other (Fig. 121). The galaxies are like surveyor's stakes which mark the distances on a construction site, not like the building bricks of a cosmic house. It is this fact which makes cosmology the study of space and time, in which galaxies provide the mass.

Galaxies are classified by their shape. The least attractive but the simplest are described as elliptical (Fig. 122). They are rounded objects, brighter in the middle than at the edges, and range in shape from apparently spherical, like an English soccer ball, to as elliptical in appearance as an American football. Unlike an American football, however, the elliptical shape probably arises from our seeing a flat disc edge-on rather than an intrinsic cigar shape. Similarly, galaxies which look circular could be spherical, or (very unlikely) cigar-shapes seen end-on, or discs seen face-on. There seem to be more circular galaxies than there ought to be if they were all discs seen face-on, so astronomers believe that many of the circular elliptical galaxies are truly spherical.

The most attractive glaxies are the spirals (Fig. 123), whose distinctive aspect is the arms, usually two, which wrap around their centres. These start outside a central region called the bulge, or nucleus, and wind out to several times the diameter of the bulge. The arms might circle the nucleus just once – loosely wrapped arms – or many times – so tightly wrapped that they may be difficult to separate. The arms lie in a flat disc, and the most photogenic spirals are those seen face-on so that the spiral arms are well displayed. The sky is full of examples of spiral galaxies seen from all angles. When a spiral galaxy is viewed edge-on, dust clouds in the plane of the galaxy are seen projected against the nuclear bulge as in M104 (p. 52), just as we see dust clouds of our Galaxy projected against the Milky Way.

Each spiral galaxy evidently rotates about an axis perpendicular to the galaxy's disc. The direction of rotation of an edge-on galaxy can be found by the Doppler shift, but the spiral arms can't be seen; the spiral arms can be seen to best advantage when the galaxy is face on and the rotation of the galaxy is across the line of sight and produces no Doppler effect. Thus astronomers have argued for many years whether all spiral arms rotate in advance of the nucleus or trail behind it.

Some galaxies, particularly the less luminous ones, have no particular shape. They are known as irregular galaxies (Fig. 124).

Spiral galaxies and populations

A spiral galaxy is a collection of stars and gas like our own Milky Way Galaxy.

As in our Galaxy these are separated in space and by age, but from our position within our own system such organisation, most evident in the colour of the components, is not easily appreciated. Only by looking at a similar galaxy from a distance does the significance of the change in colour across the disc become apparent. This clearly shows up in our

Fig. 121. *(left)* *Galaxies in space. This photograph is of a collection of galaxies in space, in fact a loose cluster where the density of galaxies is rather higher than normal. A 'normal' or average distribution on this scale would have less than one easily visible galaxy in an area as large as this. As is often the case the brightest galaxy here (IC 4329) is an elliptical.*

Fig. 122. *(above and right)* *Elliptical galaxies. These often occur in clusters, often associated with other types. The nearest large cluster of which our system may be part is shown in* (a) *(above). This is the central region of the Virgo cluster where the giant ellipticals M84 and M86 outshine all others. A more modest cluster in the southern constellation of Fornax again has ellipticals as its brightest members* (b) *(right). These are NGC 1399 (top) and NGC 1404.*

Fig. 123. *Spiral galaxies exhibit a wider variety of forms than do ellipticals, because they are intrinsically more varied and because they are discs which can be seen in any orientation.* (a) *(above) is NGC 1365, a barred spiral, a member of the Fornax cluster.* (b) *(right) is the far southern spiral NGC 6744. Both these galaxies are seen almost face-on.*

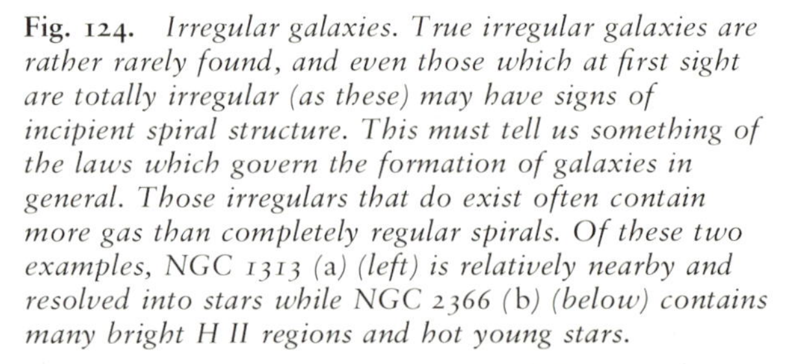

Fig. 124. *Irregular galaxies. True irregular galaxies are rather rarely found, and even those which at first sight are totally irregular (as these) may have signs of incipient spiral structure. This must tell us something of the laws which govern the formation of galaxies in general. Those irregulars that do exist often contain more gas than completely regular spirals. Of these two examples, NGC 1313 (*a*) (left) is relatively nearby and resolved into stars while NGC 2366 (*b*) (below) contains many bright H II regions and hot young stars.*

Fig. 125. *NGC 2997. Its disc is inclined at about 45° to our line of sight, revealing its internal structure and giving the galaxy an oval appearance.*

photograph of NGC 2997, a beautiful spiral galaxy (Fig. 125). The galaxy has a bright central region, or nuclear bulge, which is distinctly yellow compared with the two bluish spiral arms. The arms begin as dust lanes in the nuclear bulge and wind outwards, becoming peppered with large numbers of pinkish emission nebulae.

The distinction between the yellow bulge and blue spiral arms, so obvious on the photograph, was explained by Walter Baade in 1944 with his concept of two stellar populations. Population I is made up of the stars of the blue spiral arms and Population II is made up of stars of the reddish nuclear bulge. Yet although Baade was the first clearly to articulate this division into populations, the difference in colour between the red stars of the nuclear bulge and the rest of the spiral galaxy was first noticed by William Herschel in 1783 while observing the Andromeda Galaxy. He wrote (Herschel, 1785):

> *The ninth [nebula like our own Milky Way] is that in the girdle of Andromeda [M31], which is undoubtedly the nearest of all the great nebulae; its extent is above a degree and a half in length, and, in even one of the narrowest places, not less than 16′ in breadth. The brightest part of it approaches to the resolvable nebulosity, and begins to shew a faint red colour; which from many observations on the colour and magnitude of nebulae, I believe to be an indication that its distance in this coloured part does not exceed 2000 times the distance of Sirius. There is a very considerable, broad, pretty faint, small nebula [NGC 205]near it; my Sister [Caroline Herschel] discovered it August 27, 1783 with a Newtonian 2-feet sweeper. It shews the same faint colour with the great one, and is, no doubt, in the neighbourhood of it.*

Herschel grossly underestimated the distance of M31 (it is 250 000 times the distance of Sirius rather than 2000), but this does not detract from his amazing observation of the colour, not only of the centre of M31 (Fig. 126) but of its companion elliptical galaxy NGC 205. (This galaxy was known to Messier although not catalogued by him; Herschel is gallant but incorrect in crediting his sister with its discovery.)

From Herschel's naked-eye observations and now from colour photography we can deduce that elliptical galaxies and the central regions of spiral galaxies are redder than the arms of spiral galaxies. What is the underlying reason for this?

Fig. 126. *Andromeda Galaxy. William Miller's colour photograph of M31 was made with the Palomar Schmidt and shows the orange colour of the centre of the galaxy and its two elliptical companions NGC 205 and M32 (NGC 221).*

Populations I and II

In Chapter 4 we saw how the Hertzsprung–Russell diagram of a cluster of stars changes with time. A young cluster has a Main Sequence stretching from bright blue stars to faint red ones, while an old globular cluster has a giant branch of bright red stars and a Main Sequence of faint red ones. Spiral arms are like young clusters and elliptical galaxies are like globular clusters. How would such clusters appear from such a distance that their individual stars cannot be seen? Are the fainter stars so numerous that their combined light outshines the brighter ones? Or are the brighter ones so luminous that they outshine the far more numerous faint ones? The answer is that in young clusters the bright stars *strongly* predominate (Fig. 127) outshining by far the collective light of the fainter ones even though the faint ones are more numerous. In old clusters, like globular clusters, the brighter stars still predominate but not so strongly. In elliptical galaxies faint stars contribute far more to the collective light of the galaxy. (If collections of stars were like countries and starlight represented political power, then young clusters would have very authoritarian regimes, with a few very powerful luminaries, while globular clusters had a somewhat broader power-base. Elliptical galaxies would be more democratic, with all classes equally represented. There do not appear to be star clusters with a dictatorship of the proletariat.)

Thus a young star cluster appears blue because the brightest stars are blue; an old star cluster appears red because the brightest stars are red; and an old elliptical galaxy appears red because all the stars are red.

We ought to remember that the star colours here refer to astronomical convention rather than visual appearance. 'Red' stars have a surface temperature of 3000K on average and thus appear somewhat yellower than the sun while 'blue' ones would only seem the palest blue even if seen from nearby.

We are now in a position to explain the colours of the parts of a spiral galaxy. The central nuclear bulge is in form and colour like an elliptical galaxy or a globular cluster, and composed of old, red stars. The spiral arms contain patches, as if made up of individual star clusters, and are evidently made of very young, very blue star clusters. Somewhat older stars give the less intense, less blue colour of the smoother spiral arm structure, and the underlying disc. The star clusters have bright blue stars in them which will ionise any nearby gas; indeed they are likely to be associated with gas and dust having just formed from the interstellar medium. When the gas recombines, the strongest spectral line, the red Hα, combines with the green nebulium lines and gives a pink look to the nebulae.

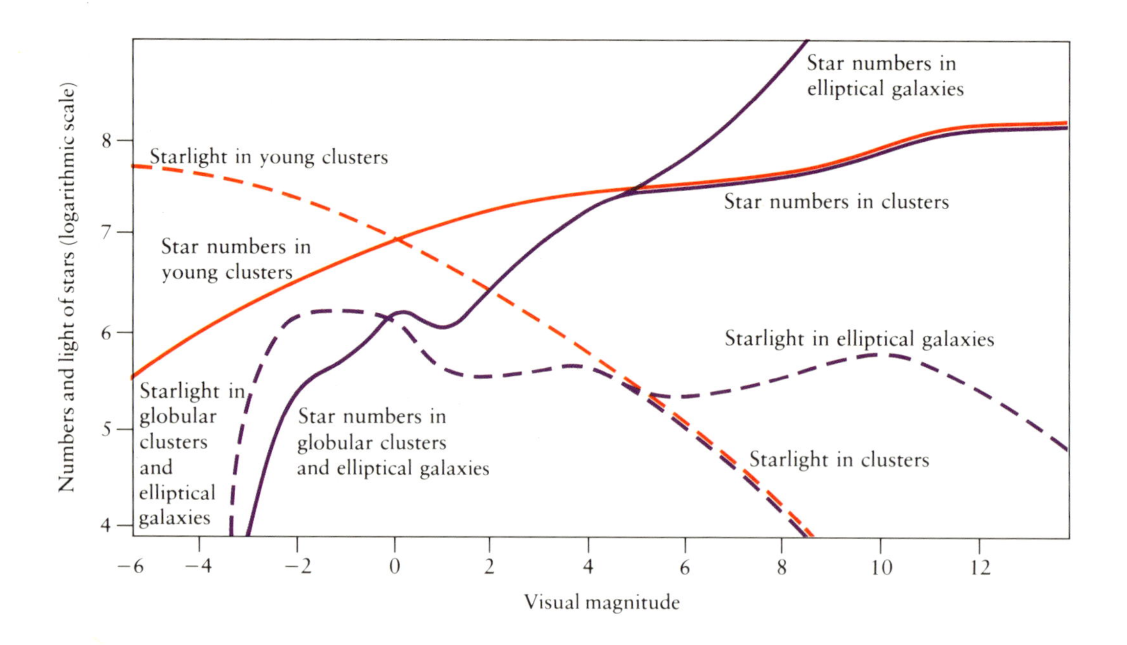

Fig. 127. *Stellar populations. The full lines represent the numbers of stars of different magnitudes in clusters and galaxies, and the dotted lines the light which they emit. Although the bright stars are always fewer than the fainter stars, in young clusters they produce most of the light. The light from the faintest stars contributes more significantly to the elliptical galaxies than to the clusters.*

In order of increasing age the coloured parts of NGC 2997 are thus: the pink nebulae, the very blue patches in the spiral arms, the less blue spiral arms, and the red stars in the nuclear bulge (Fig. 128).

M83 and NGC 253

Fig. 129 shows another spiral galaxy, M83. The less-intense coloration still shows the difference between the stars in the nucleus (red-orange) and the stars of the spiral arms (blue and patchy). M83 is a spiral galaxy, and is also known as NGC 5236. The visible galaxy appears flat, and we view it nearly full-face-on. Radio maps of M83 show that its gas extends beyond the visible spiral arms, and that it is not quite flat: one side bends up towards us and the other side curls away from us like the brim of an Australian soldier's hat.

In spite of the seemingly enormous spaces between them, most galaxies are not isolated in space. Instead they are usually found in small groups or larger clusters. The nearest group of galaxies to us is the one of which our Galaxy is a member and includes the famous spiral in Andromeda (Fig. 126) and a prominent group straddling the southern constellation of Sculptor. As in most *small* groups the brightest galaxies are all spirals.

The most distinctive of the spiral galaxies in the Sculptor Group is NGC 253, seen in Fig. 130. It appears elongated because we view it from only 17° above the rim. If we could see it face-on instead, it would appear half the diameter of the Moon. This corresponds to a diameter a little over 40 000 l.y. On this photograph, the upper left end is approaching us and the other end receding, and the galaxy is inclined so that its upper right edge is nearer to us. Dark clouds along the nearer edge are clearer because they are silhouetted against the stars of the bluish disc of the galaxy which lies beyond. A few, bright red HII regions are visible and these seem to be distributed at random, instead of being closely associated with the spiral arms as was the case in NGC 2997, though this might be a function of the angle at which we see the galaxy, rather than a real effect.

The B, V and R plates used to make this colour picture were taken on the AAT by Ken Freeman long before our colour photography project originated and for quite a different purpose. Their usefulness indicates the potential of archive material with this colour process.

There is, however, a good deal more to NGC 253 than is seen in the colour picture. The photographic amplification technique described in Chapter 2 claimed this galaxy as one of its first successes, revealing the very large, unstructured halo shown in Fig. 131. It has been known for some time that some spiral galaxies have large, low-surface-brightness envelopes beyond the usually accepted limit of the photographic image, but that associated with NGC 253 is odd in at least two respects. It is asymmetrical, showing a distinct bulge to the south, and it is considerably elongated whereas such haloes are more often described as ellipsoidal.

Fig. 128. *Spiral arms. In NGC 2997 the pink gaseous emission nebulae and the dark patches of dust associated with them lie near the blue star clusters on the inside edges of the spiral arms. It can be seen that the young material lies on the inside edge of this spiral which thus leads the rotation of the galaxy, unlike, say a Catherine Wheel or water in a whirlpool, where the spiral trails back from the direction of rotation.*

Fig. 129. *(left)* M83. *This picture of a face-on spiral galaxy was made of Kodak Vericolor II negative film, rather than by the three-colour superimposition technique and shows much less intense but similar colours to the picture of* NGC 2997.

Fig. 130. *(above)* NGC 253. *This edge-on spiral galaxy, is one of the dustiest known, and much of its internal detail is hidden. However two spiral arms composed of bluish clusters of stars can be seen in the outer parts of the galaxy. The upper (north-western) edge is nearest to us.*

In the absence of any obvious interacting companion one might examine the structure of the underlying spiral to see if it is disturbed. A close look at Fig. 132 shows that the southern half of the galaxy is rather like the northern, with no systematic disturbance large enough to account for the bulge.

Elliptical galaxies

Colour photos of elliptical galaxies are rare, because they are generally not as picturesque or interesting as spiral galaxies. Fig. 133 shows NGC 5128 or Centaurus A, which is an atypical elliptical galaxy but shows the elliptical shape and colour. Had Charles Messier, the eighteenth-century cataloguer of nebulae, lived farther south, NGC 5128 would have had an M-number like so many other prominent galaxies, but it lies in the southern constellation of Centaurus, and is difficult to explore from European observatories.

The bright galaxy's title of Centaurus A results from it being the most intense radio source in that constellation. When pinpointed, the radio source was identified with NGC 5128. At that time NGC 5128 was thought to be a peculiar nebula within our own Galaxy. Only when Palomar astronomers Walter Baade and Rudolph Minkowski studied NGC 5128 was its extra-galactic nature revealed.

Fig. 133 shows the principal features. Most of the light comes from an almost spherical collection of yellow Population II stars which is classified as a giant elliptical galaxy. The galaxy is slightly elongated along a north–south line. Across its waist, trending east–west, runs a dark lane of dust and gas, intricately woven into streaks and bands. Baade and Minkowski saw this dark band as the arms of an otherwise unseen spiral galaxy, and believed Centaurus A a collision between an elliptical and a spiral galaxy. The energy generated by such a collision could be released as radio waves, they argued in 1954. The idea that the collision itself produced radio waves was soon discredited and this led to a reaction against the idea that this object was a collision of two objects. Instead, astronomers concentrated on the concept that the band of dark material was a girdle which completely encircles

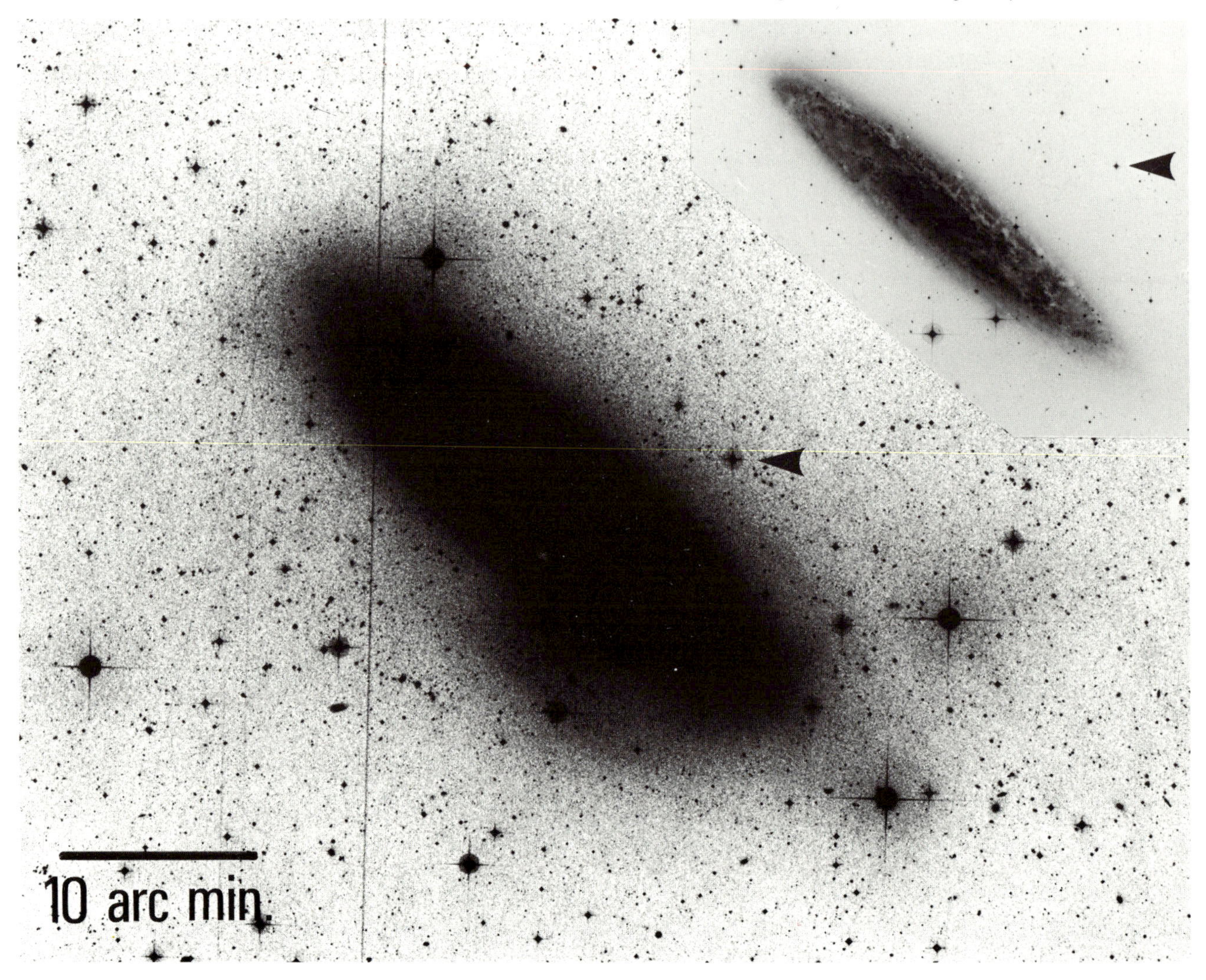

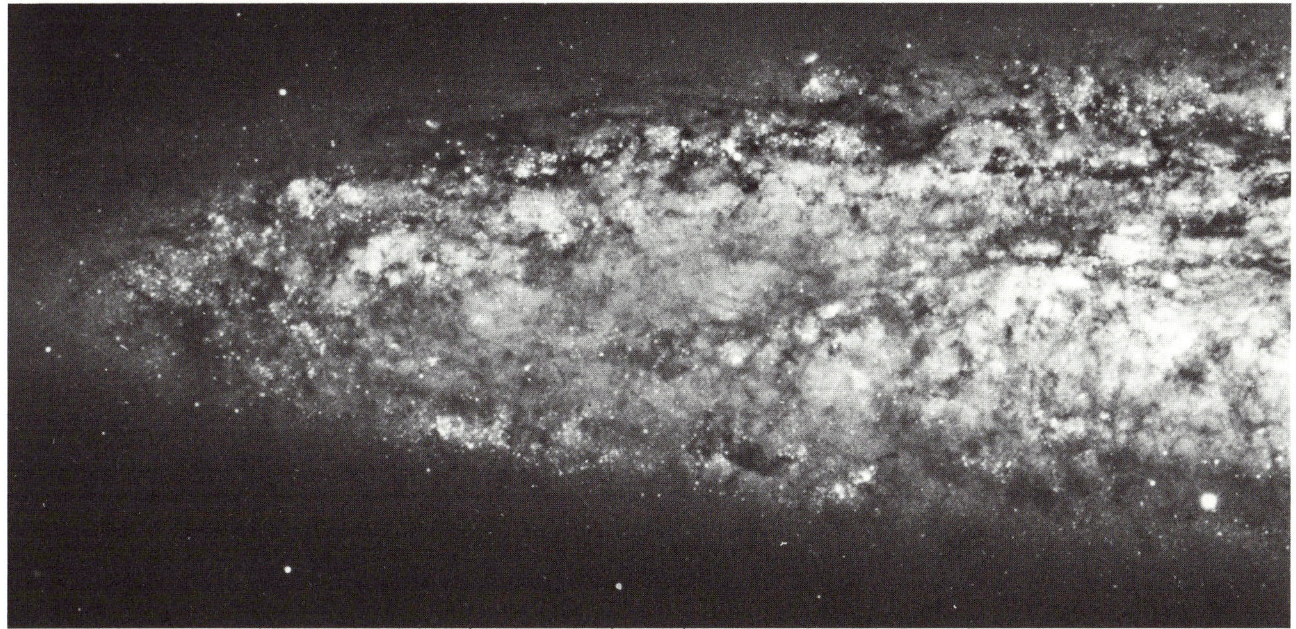

Fig. 131. *(left)* *NGC 253 amplified. A photographically amplified image occupies the main part of this picture; inset is a representation of the original plate. Both pictures were made from the same plate and are printed to the same scale. The same star is arrowed in both to indicate the extent of the low-surface-brightness envelope around the galaxy. Note the large protrusion to the south.*

Fig. 132. *(above)* *The stars of NGC 253. Each extremity of NGC 253 is seen here partly resolved into stars and nebulae. This beautiful deep plate was taken by Dave Hanes on the AAT. The pictures reveal extensive small-scale structure. The two extremities are remarkably similar, except that each is a mirror image of the other.*

NGC 5128 and that it was a single galaxy. They regarded the unusual appearance of the dust in an elliptical galaxy as a failure of whatever mechanism it was which removed dust from elliptical galaxies. Astronomers began to regard elliptical galaxies as spirals like M31, which had passed through one another clearing out the debris and leaving the stars behind. The stars aged and reddened, and could not be replaced by young stars because there was little interstellar material to produce them. NGC 5128 is an isolated galaxy – perhaps in its long life it had never encountered another and the dust had remained. In recent years John Graham, using data from the 4-m

Fig. 133. *Centaurus A. Basically an elliptical galaxy with a population of yellowish stars, Cen A is crossed by a band of dust which reddens the stars still further. Formed from the dusty material, blue stars and pink H II regions can be perceived along the rim of the dust band.*

Fig. 134. *Centaurus A masked. The elliptical galaxy has almost disappeared from this photograph made with an unsharp mask. Blue stars and pink emission nebulae inhabit the whole dusty area, and the orange nucleus of the galaxy is revealed at its centre.*

Cerro Tololo Telescope in Chile, has argued, however, that NGC 5128 is a merger of two objects. In 1979 he concluded that 'the structural peculiarities of NGC 5128 . . . suggest some sort of massive addition of gaseous material to a basically normal elliptical galaxy in the not too distant past'. The origin of the gaseous material is not clear, but perhaps NGC 5128 merged with an intergalactic hydrogen cloud or young dusty galaxy during a chance encounter. Thus, although basically an elliptical galaxy, NGC 5128 has a disc comprising gas and dust, and maybe some stars; it therefore has some affiliations with spiral galaxies.

The colour photo in Fig. 134 was produced from the same three plates as was Fig. 133, but masked to suppress the elliptical galaxy. It shows that the blue stars on the edge of the dust lanes and pink emission nebulae of Population I material completely surround the dark dust lane.

Radio maps of Centaurus A show it to be extremely large, fully $6^\circ \times 10^\circ$ in the sky. The principal areas of radio emission are two lobes extending roughly to north and south, and lying at right angles to the dark girdle. In Centaurus A, therefore, we have evidence that the radio lobes lie at the poles of the galaxy. Moreover, it was shown in 1961 that Centaurus A is a double-double radio source. A second pair of radio sources lies on the same line as the large lobes: these are much smaller, and lie near the northern and southern edges of the visible distribution of stars. It seems that two explosions have occurred at the centre of NGC 5128, each ejecting a pair of radio sources out of the galaxy's poles.

The galaxy generates optical emission in the direction of the radio lobes. Fig. 135 shows faint projections extending up to one degree from the centre of the burnt-out image of the Centaurus A elliptical galaxy. The cause of these projections and their relation to the radio lobes are unknown.

These pieces of evidence focus our attention on the very heart of the galaxy whence the explosions might have originated; unfortunately, this lies behind the dark dust lane and is thus difficult to study. The effect of the dust, like smoke from a chimney, is to redden the light from the nucleus behind it. A deep orange spot of light can be seen at the very centre of the galaxy, and is visible in Fig. 134. This is a giant cluster of hot stars and gas with a total luminosity 2 billion times greater than that of our Sun, and thus brighter than many entire galaxies. Even this, however, is not the true heart of NGC 5128. At the centre of this cluster of stars is a compact source of radio, infrared and X-radiation. Here lies a dense and very luminous object, the true nucleus of the galaxy. The X-rays show it to be a variable object, with flares. One speculation is that these flares represent individual large stars being swallowed by a black hole.

Fig. 135. *Centaurus A amplified. Centaurus A has recently been shown to be much larger than previously suspected. This picture shows that the outer envelope is considerably extended at right angles to the dust lane and roughly along the axis of the radio lobes. Inset is a more usual image of the galaxy, derived from the same plate and printed to the same scale.*

NGC 6822

Centaurus A is a huge and unusual galaxy but NGC 6822 is small and ordinary. Fig. 136 pictures a galaxy in which individual stars are easily photographed and bright clouds of glowing gas can be recognised. One might even suspect that the stars of this galaxy could be counted, but a photograph such as this reveals only the very brightest specimens, and the stars probably number hundreds of millions.

Individual stars can be seen not merely because they are isolated from their neighbours, but also because they appear bright. This is not a peculiarity of the stars themselves, but an effect of the proximity of NGC 6822. Lying at a distance of about 2 million l.y., this is one of the closest galaxies to us, and a member of our Local Group. The brightest stars give out 500 000 times as much light as our Sun, and are comparable to the brightest stars in most larger galaxies.

No symmetrical pattern exists in NGC 6822: the galaxy is irregular, and about 10 000 l.y. across. There is a prominent bar of bright nebulae and stars across its top in Fig. 136. A curving band of neutral hydrogen gas extends from each end of the bar of nebulae. The band resembles loose spiral arms. Possibly a photograph taken 100 million years hence, when the gas in these arms has condensed to form stars, would show NGC 6822 to be a small spiral galaxy.

H II regions in our Galaxy comprise about 93% hydrogen, about 7% helium and only traces of all other elements. The 15 H II regions in the bar of NGC 6822 contain even smaller proportions of the other elements. This indicates that the galaxy is immature, for hydrogen and helium are only very gradually converted by stars into heavier elements. The general mixture of dust, gas and blue stars in NGC 6822 resembles that in the spiral arms of NGC 2997 (Fig. 128).

The Magellanic Clouds

The Magellanic Clouds are named after the Spanish explorer Magellan whose name became associated with the two galaxies after the publication of the record of his ship's round-the-world trip of 1518–20. The Clouds look like two fragments, large and small, which have broken off the Milky Way but they are two individual galaxies, in fact the nearest to us. Apart from our own Milky Way, these two rather substantial fragments are the only galaxies readily seen without a

Fig. 136. *(above)* NGC 6822. *Individual stars and bright pink nebulae can be discerned in this nearby galaxy. As in the Large Magellanic Cloud, most of the star formation seems to be concentrated at one end of the central bar.*

Fig. 137. *(right)* *Small Magellanic Cloud. An irregular galaxy, the SMC contains numerous blue stars; some pink emission nebulae are visible to the left of the picture, but there are less than in the LMC.*

telescope but are not visible from countries well north of the equator.

The Large Magellanic Cloud (LMC) is only 160 000 l.y. away, the Small Magellanic Cloud (SMC) somewhat further. Indeed the two galaxies are orbitting our own Galaxy at a distance of only three of its diameters or so. The colour picture (Fig. 137) of the SMC shows a multitude of blue stars and a few pink nebulae in an irregular form, like NGC 6822. Also like NGC 6822, the SMC is very deficient in the elements heavier than hydrogen and helium.

The LMC is about 30% nearer to us than is the SMC and seems larger partly because of this; however, it really *is* larger, with a greater number and variety of stars and nebulae (Fig. 138). The same Population I mixture of bright blue stars and pink nebulae predominates, but the form of the galaxy is not quite irregular. The nebulae extend in opposite directions from the end of a huge bar of stars and nebulae, like an incipient spiral. The bar of the LMC indeed shows increased numbers of yellow stars in the central regions, compared with the SMC. The evidence

Fig. 138. *(above) Large Magellanic Cloud. The band of stars stretching across the picture is the bar of a spiral galaxy, with emission nebulae at the left end turning along a vestigial spiral arm. The brightest nebula is the Tarantula Nebula, seen enlarged opposite; the nebulae at the top of the picture are shown enlarged over the page. Some of the brightest stars in this picture, in the foreground, belong to our Galaxy, but the vast majority of the stars visible belong to the Large Magellanic Cloud. They remain individually distinguishable. Thus the LMC has been important to the study of astronomy, giving an insight into how parts of another galaxy compare with similar parts of our own. Figs 139 and 140 show nebulae which can be studied with almost as much detail as run-of-the-mill nebulae in our own Milky Way.*

Fig. 139. *(right) Tarantula Nebula. Also known as the 30 Doradus nebula, the Tarantula is the brightest H II region at the top of the row of nebulae forming one spiral arm of the LMC.*

Fig. 140. *(pp. 174–5) Large Magellanic Cloud. Numerous H II regions and bright young stars (including N70 in lower left, see p. 124) dominate this northern region of the LMC.*

of shape and colour marks the LMC as a barred spiral galaxy with a nuclear bulge.

Offset from the centre of symmetry of the LMC is a bright H II region, just visible to the unaided eye, the Tarantula Nebula (Fig. 139). This is the only gaseous nebula visible in an external galaxy without a substantial telescope. Its naked-eye appearance led to its being given the name of a star, 30 Doradus. The mixture of very old Population II stars and very young Population I objects found here (including at least one star which may have a mass of several hundred Suns) marks this as the probable nucleus of the LMC, off-centre or not. This spectacular H II region is extremely large, about 900 l.y. across.

To the north of 30 Doradus is the amazing collection of H II regions and young bright stars, shown in Fig. 140, a continuation of one of the spiral arms of the LMC.

NGC 4027

Fig. 141 evidently shows a galaxy of a blue colour, with the implication that it is Population I material; and the fact that it has two arms show it to be a spiral, even if distorted. The motion within the hook of the main arm of the galaxy as revealed by the Doppler shift from the individual H II regions is like a whirlpool or eddy within the main rotation (de Vaucouleurs, de Vaucouleurs and Freeman 1968). NGC 4027 is a relatively low-mass galaxy with a pronounced bar across the centre. Although the LMC does not have arms as pronounced as this example, NGC 4027 is in fact a small galaxy, comparable to the Large Magellanic Cloud in size and shape.

Fig. 141. NGC 4027. *The main spiral arm hooks from the far side of the galaxy and points towards us while a stubby spiral arm starts on our side of the galaxy and projects across the line of sight. Numerous pink H II regions can be seen on the outside curve of the stubby spiral arm, associated with streaks and patches of dust silhouetted against the galaxy. More H II regions are visible over the galaxy on the far side of the main spiral arm, and on its inside curve, including one right at its tip.*

Chapter Nine

Red-shift, blue-shift

Doppler effect

It is a common fact of our experience that the pitches of moving sources of sound (train whistles, ambulance bells, police sirens, car engines) change as the sources move relative to us. The reason is that the sound waves emitted by the whistle, bell or whatever, are crowded together in the direction of motion of the source and stretched in the direction away from the motion. Thus the perceived wavelength of sound waves changes and the pitch of the sound is different. This is called the Doppler effect, after Christian Doppler who formulated the principle in 1842. His idea has wide and important applications, although it took a long time to be recognised as important, perhaps because Doppler held relatively minor positions during a working life in Prague and Schemnitz and was relatively briefly a professor in the University of Vienna, dying at 48 of lung trouble.

Wavelength is an attribute of light as well as of sound, and wavelength is related to colour. Does this mean that moving sources of light change colour? Doppler thought so, and speculated about a possible example. He was wrong about the particular example, but right in principle.

Colours of moving stars

Doppler was on firm ground in using the Doppler effect to explain the apparent change in pitch of moving sources of sound. The wavelength of the sound wave increases as the source moves away from the observer (lower pitch).

The first experimental verification of the Doppler effect, by Ballot in 1845, used whistles on a moving train. He measured their frequency by using musicians with the gift of absolute pitch. A very noticeable change in pitch by one note in an eight-note octave is a change of frequency by a factor of $2^{1/8}$, and corresponds to a speed of 9% of the speed of sound, which is 330 m/s in dry air. Speeds in the region of 100 km/h are not uncommon to our experience, so that we are familiar with the Doppler effect changing the pitch of train whistles, ambulance bells, police sirens and car engines.

Doppler realised that his principle could be applied to moving sources of light. Indeed, as early as 1676 Roemer pre-empted Doppler by using the apparent change in orbital period of Jupiter's satellites to determine the speed of light. He did not generalise the principle behind his experiment and thus lost the chance to have the Doppler effect known as 'Roemer's principle'. Searching for examples of the effect in light waves, Doppler attributed the different colours of stars to differences in their presumed speeds. In his deductions he made a colossal blunder.

He knew that the spectrum of light consisted of the colours from violet to red but apparently did not know of or guess the existence of the radiations beyond the red and violet, the infrared and the ultraviolet. He thought that the spectrum of a star would terminate at the edges of the visible spectrum. If a star at rest emitted white light of all colours, then if it receded fast enough the violet of its spectrum would be displaced to the blue, the blue to the green and so on, leaving a gap where the violet was. A further increase of speed would leave a larger gap. Thus the only visible parts of the spectrum would be the redder ones and the star would be coloured red. Conversely, an approaching star would appear blue.

This idea was generally greeted with scepticism. In the same paper in which Ballot gave in 1845 his experimental confirmation of the Doppler effect in sound he scorned the idea. William Huggins in 1868 dismissed it with the remark:

> *That Doppler was not correct in making this application of his theory is obvious from the consideration that even if a star could be conceived to be moving with a velocity sufficient to alter its colour sensibly to the eye, still no change of colour would be perceived, for the reason that beyond the visible spectrum, at both extremities, there exists a store of invisible waves which would be at the same time exalted or degraded into visibility, to take the place of the waves which had been raised or lowered in (wavelength) by the star's motion.*

What misled Doppler was not only his ignorance of the existence of infrared or ultraviolet radiation, but a set of spurious evidence about star colours which apparently supported his correlation of star colour with velocity. For instance, he says that when double stars are coloured, if one is much brighter and hence presumably much more massive than the other, it is white, and the fainter is coloured, because it is moving in orbit about the other which is almost stationary; but when they are of equal brightness one is red and the other blue because each moves about the common centre of gravity, one receding, the other approaching. If these statements are true (and a glance at the selected coloured stars in Fig. 16 does not readily confirm them), then they may be psychological contrast effects. Doppler says that in the 50 years between William Herschel's observations of the colours of Gamma Leo and Gamma Del and those of F.G.W. Struve, the colours have changed, as a result of orbital motion and a change of radial velocity. The range of estimates of the same star by the two men is very wide and such reported changes are of low reliability, because of contrast effects, dark adaption, the Purkinje effect, etc.

A series of observations by the same observer with the same instrument of the same star could be more reliable and the German astronomer Hermann J. Klein (1867, 1876) published two series of observations of the colour of Alpha Ursa Major. His papers show a meticulously carried out, systematic investigation of colour, on a carefully graded scale, in which Alpha UMa varied from 'chrome yellow' through 'reddish yellow' and 'yellowish red' to a 'pale firey red'. This is equivalent to a change in colour index ($B-V$) from about 0.6 to about 1.8, according to Minnaert's observations, and should be clearly real. The colour changes had a period of 35 days.

The star is indeed known to be a binary star – but with a period of 44 years. Far from having a variable colour and magnitude, it is one of the standard stars of the UBV magnitude system, that is, it has been chosen as a reference star, relative to which other stars may be measured, and is constant to the much more precise photometers available nowadays. Why Klein's careful series of observations showed such a resoundingly false result is a mystery, but the story illustrates the difficulty in evaluating visual evidence about star colours and perhaps we should forgive Doppler for being misled by the spuriously precise impression of colour which had led some astronomers to believe that they had seen changes in the colours of stars. As we have seen in Chapter 1 the psychological circumstances in seeing colour in faint star images are rather confusing. Some astronomers find it difficult to see colour at all. Contrast, dark adaption, the Purkinje effect, all these and numerous other problems make the estimation of colour difficult. Tyndall, describing the Doppler effect for sound in 1867, summarised his view of the Doppler theory of star colours: 'The ingenuity of the theory is extreme but its correctness is more than doubtful'.

Of course it is now known that the colours of stars predominantly indicate not their different speeds but their different temperatures; and we know that the spectra associated with these temperatures have peaks at certain wavelengths. Thus the motion of the star, which will change the peak wavelength, does alter its colour.

A Doppler-shifted black body of a certain temperature has the identical spectrum to a black body of another temperature. If a star like the Sun is receding at a speed of about 10 km/s, its apparent temperature decreases by about 0.5K. This small speed produces a small change of temperature and a slight, but completely imperceptible reddening of the star's colour.

Speeds of about 10 km/s are sufficiently large to produce a measurable shift in the position of the Fraunhofer lines in the stars' spectra, and indeed this is the way that the radial velocities of stars are measured. A 10 km/s velocity produces a wavelength shift of about 0.1 Å in the position of a spectral line but this is far smaller than the 20 Å or so shift which our eyes require to perceive a colour change.

High-speed changes

Doppler himself realised that very large speeds would be required to produce a change in colour of the stars. We recognise in the spectrum of white light a rainbow of seven colours – red, orange, yellow, green, blue, indigo, violet – and these span, as it happens, about one octave in frequency of electromagnetic radiation, so that a very noticeable change in colour is produced by a speed of about the same fraction of the speed of light, as a change by one note in an eight-note octave. If Doppler were correct then the stars must have radial speeds of about one tenth the speed of light (*c*; which is 300 000 km/s or 186 000 miles/s). This is several thousand times the actual speeds of stars. Doppler should have realised the conflict between his hypothesis and the then current measurements of the angular speeds and distances of nearby stars. In 1 year a star travelling at a tenth the speed of light travels 0.1 l.y. If it is at a distance of 10 l.y. it moves through an angle of 0.01 radians (rad). Its angular speed is thus 0.01 rad/yr (0.5°/yr). Nearby stars would move a distance on the sky equal to the Moon's diameter in a year or so. The observed angular speeds of nearby stars are nowhere near this size, being at most 10 arc seconds/yr.

Thus, in spite of the scorn of his contemporaries, Doppler was in principle correct that a radial velocity of a source of light would produce a change in its colour, but only when the velocities concerned are comparable with the speed of light can a change in colour be perceived, and even the subject matter of astronomy did not usually produce such extremes. Nevertheless there are two modern discoveries which demonstrate the change in colour by radial velocity: the red-shift of distant galaxies, and the bizarre case of SS433.

The K-correction

Until 1979, only in external galaxies had the change in colour produced by radial velocity been seen. The Doppler shifts produced in moving stars are small, but galaxies mark the expansion of the Universe. In 1936 Edwin Hubble showed that all galaxies were receding from Earth and that the more distant ones were receding faster, with a red-shifted spectrum. In fact, the most distant detectable ones recede from us at speeds up to several tenths of the velocity of light. The spectrum of a galaxy is essentially the same as the spectrum of the stars which it contains. As the galaxy recedes, the temperature of the stars seems cooler, and so they seem redder. This measurable effect is part of a calculation called the K-correction. To put the explanation of the effect another way, the ultraviolet light emitted by a distant galaxy receding from us is red-shifted into the visible region of the spectrum. If there were a relative lack of ultraviolet light in the spectrum of the distant galaxy it would be interpreted by us as a deficiency of blue light, which would make the galaxy seem redder.

First attempts systematically to calculate and measure the K-correction were made by Humason, Mayall and Sandage in 1956, and showed a *reddening* of all galaxies; but astronomers did not expect that more distant galaxies would actually show a *bluening* due to larger velocities. This is because, until the launch of the Orbiting Astronomical Observatory-2, which was a satellite for ultraviolet astronomy, astronomers had little knowledge of the ultraviolet emission from galaxies. It turned out that elliptical galaxies do appear generally redder as we look at the more distant ones. Spiral galaxies begin in the same way, becoming redder by about two steps of Minnaert's colour scale, but they contain such large numbers of hot ultraviolet emitting stars (Population I material). Moreover, when looking back at distant galaxies, we see them earlier in the history of the Universe, when, apparently, larger numbers of Population I stars were being formed. These stars are hot and bright. When the ultraviolet emission from them is shifted into the visible spectrum, very distant spiral galaxies actually become bluer-looking!

It is as if a receding vehicle carried not only a horn whose note was made deeper by the Doppler effect of recession, but also a loud ultrasonic dog whistle which became audible and drowned the horn.

These large Doppler effects which changed the colour of receding galaxies were not paralleled by any galactic object until 1979. The object which changed this is called SS433.

SS433

SS433 is the 433rd entry in a catalogue by N. Sanduleak and C. Stephenson of stars having emission-line spectra. Such stars have a spectrum consisting of an underlying essentially black body component – the optically thick surface layer of a star – plus emission lines coming from an optically thin gas surrounding the star. The stars are catalogue-worthy because the gas represents some interesting interaction between the star and its neighbourhood.

The interaction causing an emission-line spectrum may have any of many causes. Some of the stars in the SS catalogue are rotating so rapidly that centrifugal force at their equators counteracts their attractive gravitational force, so that the stars shed an equatorial disc of gas which shows as the emission-line region. Some, like SS433, are binary stars with matter flowing around and between the two stars. Since hydrogen is the most common element in the Universe, the stars in the SS catalogue typically show hydrogen emission lines (the Balmer series, the strongest line of which is Hα at 6563 Å). Helium lines (from the second most common element) are also typical.

SS433 languished as one star among many hundreds in the SS catalogue until its rediscovery in 1978. Spectra taken then show the Balmer series and spectral lines from neutral helium, both arising from the circumstellar material ripped from one star by its companion. There were also present in its spectrum other lines which were unidentified, particularly because they seemed to come and go sporadically.

It was only after several months of monitoring the spectrum of SS433 that the pattern became apparent (Fig. 142). The unidentified lines appear in pairs, the most prominent of which shift cyclically in position about a wavelength which is somewhat to the red of the Hα line. The cycle has a period of 164 days, and the range over which each member of the pair shifts is more than 1000 Å. The lines are antiphased. With this clue to the interpretation of the two strongest of the unidentified lines, the other unidentified lines could then be paired off. They show similar behaviour, oscillating in antiphase about wavelengths which lie to the red of the other Balmer lines and the neutral helium lines.

Thus the emission line spectrum of SS433 consists of three components – a stationary set of Balmer and helium emission lines arising from the interaction of a double star, and two antiphased oscillating set of moving spectral features associated with the Balmer and helium emission, and arising from a new phenomenon. The wavelength shifts are so large that the bluer of the moving Hα features ranges in colour from yellow (its minimum wavelength closely matches the sodium D-lines near 5900 Å) through

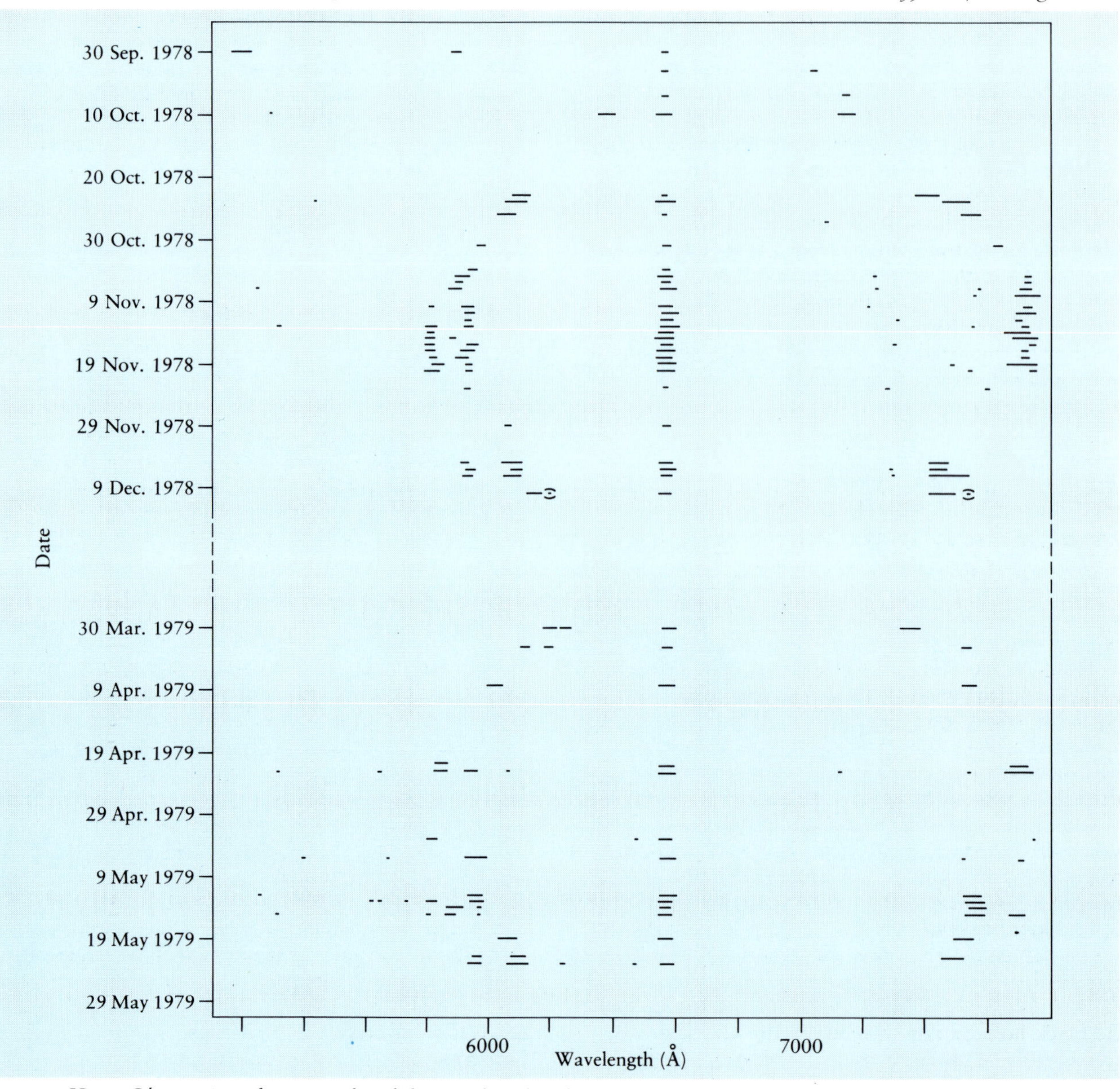

Fig. 142. *SS433. Observations for 9 months of the wavelengths of the emission lines in the red and yellow regions of SS433 (the gap in coverage occurred when the Sun was in front of SS433 and it was therefore not visible at night). The stationary Hα emission line occurs at a constant wavelength of 6563 Å. Either side of Hα are two emission lines in antiphased cycles. Other emission lines in the spectrum slightly confuse the simple picture. Data from A. Mammano, F. Ciatti and A. Vitalone, Asiago Observatory.*

orange, and the deep red of the stationary Hα line, to nearly 7000 Å, which is beyond the red response of the eye – infrared in fact. Even if he was wrong about the colour changes in double stars, Doppler's speculation of 140 years ago attributing colour changes to changes of speed had found a problem to solve in 1979.

The current interpretation of the moving features does not use orbital motion, but is in terms of a pair of equal and opposite jets of material shooting from SS433. Because of the speed of outflow of material, the spectral emissions of hydrogen and helium gases in each jet are Doppler-shifted from their rest wavelengths, one (the approaching jet) generally to the blue and the other (the receding jet) generally to the red. The jets precess like a spinning top in a conical motion with a period of 164 days and so the features move in antiphased cycles of this period.

Curiously, even at the moments when the jets are in the plane of the sky, with no component of speed towards or away from us, there is a shift. It has its origin in the phenomenon of relativistic time dilation. The fast-moving hydrogen atoms in the jets have 'clocks' – the natural time scales of atomic phenomena – which are running slow with respect to ours and so emit Hα photons of lower frequency. The speed of the material in the jets is an astonishing one quarter of the speed of light!

The jets of SS433 in space

Interest was focused on SS433 because it was identified with an X-ray star within a spherical radio shell called W50. W50 is a supernova remnant – an expanding spherical shell produced in the interstellar gas by a long-ago supernova explosition – and SS433 lies plumb at W50's centre. This led to the suggestion that SS433 is the stellar remnant of the same supernova explosion which gave rise to W50. Perhaps SS433 contained a black hole produced in a supernova explosion on one of a pair of stars. In this model, the black hole accretes material expelled by its companion (this material gives rise to the stationary emission lines which earned SS433 its place in the SS catalogue). As it approaches the boundary (event horizon) of the black hole, in-falling material is compressed by the intense gravitational field of the black hole. It is heated to billion degree temperatures and yields X-rays which may be the reason why SS433 is an X-ray star. (Incidentally, the accretion may be at such a rate that the black hole cannot swallow all the material. The pressure of the radiation generated by heating the material may be the power-house driving the relativistic jets.)

Not only is SS433 an X-ray star, it is also a radio star – indeed, not knowing of its SS designation, the rediscoverers of SS433 used the *radio* co-ordinates to point the 3.9-m Anglo-Australian Telescope to the *optical* star which they anticipated was the stellar remnant of the supernova. Looking at the radio star with the Very Large Array of radio telescopes in New Mexico has resolved the radio image of SS433 into a blobby, elongated structure in which the blobs move out in a cone, with transverse speeds of 0.0088 arc seconds/day – matching the 0.26*c* jet speed at the 16 000 l.y. distance of the star. More than this, the directions in which the jets point – roughly east–west on the sky – align with two 'blisters' or 'ears' in the radio shell of W50 which are presumably created by the pressure of the jets on the inside of the supernova remnant. Thus the radio observations confirm the deductions of optical astronomers about the existence of the precessing jets which they invented to explain the Doppler shifts in the spectra of SS433.

No doubt, Doppler would have been gratified to see the colour changes in the spectrum of SS433, following his fruitless speculations of 140 years earlier.

Future celestial images

In this book we have concentrated only on the celestial colours which can be seen or recorded by astro-photography. We have shown 'real' colour pictures. We have allowed ourselves to enhance in these photographs the colours which exist in the sky: we have used photographic techniques like unsharp masking, contrast enhancement and three-colour addition, just as children might play with the controls of a television receiver in order to 'turn up the colour'. Underlying our exaggerations are the colours of reality as it might be perceived.

What happens in the future? What is the future path of coloured celestial imagery? In a sense the future is already upon us. Any image can nowadays be represented in colour via a computer. It can be read into the memory of the computer as entries in a two-dimensional array, like the spot-heights on a surveyor's map. A look-up table in the computer programme can convert the entries into colours, for instance coding all areas on the map with heights less than 100 m above sea level as green, and all heights over 200 m as white, with shades of brown and purple in between – just like a map in an atlas, in fact. The entries in the array can then be fed as signals to the B, G and R inputs of a colour monitor and displayed as a coloured image. The image need bear no relation at all to the 'real' colours of the object portrayed. In astronomy, images created by radio-telescopes, for example, can be colour-coded such that bright areas of intense radio emission appear red and faint areas green, with oranges and yellows in between. This adds drama to the image and, with a careful choice of the look-up table, can bring out relations between parts of the map which are otherwise unclear. The range of possible techniques in astronomy is limitless, and many are illustrated in Nigel Henbest and Michael Marten's book, *The New Astronomy* (CUP, 1983). These are often not, from a common-sense point of view, images that you might really have seen. On the other hand, radio maps of galaxies can be colour-coded to show their motions, with red for the receding bits and blue for the approaching bits, or infrared maps of nebulae can be colour-coded to show their temperature. Both these examples translate a kind of colour at another wavelength into our kind of visible colour. In a way, this just extends our vision to radio or infrared wavelengths, and shows what we might have perceived if our eyes were sensitive to these radiations (Fig. 143).

What extensions to human vision are left in 'real' colour imagery in astronomy? New detectors of light, such as Charge Coupled Devices (CCD), convert light into electrical impulses, ready to be fed directly to a computer-like memory. These detectors will extend the sensitivity of the human eye. The Space Telescope, to be launched at the end of the 1980s will have electronic detectors like this, working above the Earth's atmosphere, recording images from the clarity of space. In these conditions it might be possible, for instance, to record the jets of SS433 emanating from the bright central star. Then we will see the jets oscillating from side to side, changing colour from yellow to deepest red, and back, as they precess, in an astronomical colour movie. Deep exposures with the Space Telescope, recording galaxies at lookback times of 10 000 million years, could be made into colour images showing numerous blue galaxies undergoing the burst of star formation which apparently occurred soon after the Big Bang. It is 100 years since Ainslee Common produced the first deep astrophotographs – the indistinct, grainy, black-and-white photographs of Chapter 2, which started a trend leading to the fine-grain, deep, colour pictures in the rest of this book. In the next 100 years the primitive electronic images available to us now will undergo a similar evolution.

Fig. 143. *Orion in the infrared. Data from the Infra-Red Astronomy Satellite (IRAS) has been processed into a colour image of Orion as it might be perceived by a being with infrared-sensitive eyes. Red colour represents strong 100-μm radiation, green strong 60-μm radiation and blue strong 12-μm radiation, so that the red-to-blue sequence represents a temperature increase as in star colours. The Rosette Nebula is the bright yellow patch centre left and the Orion Nebula (M42) is the bright nebula below centre, with the Horsehead Nebula above it. The large ring is centred on Lambda Orionis and is the heated dust shell surrounding the Lambda Orionis H II region. Compare this view of Orion with Fig. 90.*

Appendix I: Image manipulation

Introduction

Most of the photographs of astronomical objects which appear in this book have been derived from plates taken on the Anglo-Australian Telescope (AAT) or on the UK Schmidt, both located on Siding Spring Mountain in north-western New South Wales.

Both these telescopes began operation in 1974–75 and were among the first to exploit the fine-grain IIIa emulsions described in Chapter 2. The ability to use these relatively slow, high-contrast plates is due in large part to effective hypersensitisation procedures; the UK Schmidt (see Sim, Hawarden and Cannon, 1976) were among the first to incorporate hydrogen gas as suggested by Babcock *et al.* (1974) into a practical working system. Deep plates are generally exposed to achieve optimum detection of faint objects (i.e. maximum output–signal-to-noise) and have minimum density of 1.3 and a maximum often exceeding 4.0 (ANSI diffuse densities). Such dense, contrasty negatives are essential in astronomy, but demand special techniques to extract the full range of image information from them.

This section describes the materials and processes behind these methods and explains how the enhancement techniques designed initially for monochrome photography are incorporated into procedures for making colour pictures.

Over the years the processes have been steadily refined and simplified and can be summed up in three main principles:

1. Original plates are never subject to further chemical processing. All image manipulations are performed by contact copying and the final result is obtained either by contact or enlargement of some derivative.
2. All contact printing is done with a diffuse light copier.
3. Only three types of film are used in the monochrome stages and all these films are developed in the same developer.

All the materials used are commercially available and the developer is a concentrated liquid for easy dilution. It should be emphasised that these procedures were developed in Australia where photographic suppliers import only a limited range of products. This accounts in part for the use of some materials in a way their manufacturers never intended.

Diffuse-light contact copying

It was shown by Malin (1977) that a large-source, diffuse-light contact printer is capable of extracting image information from the extremely high densities found in thick, fine-grain emulsions after vigorous development. Such a source illuminates an original film or plate over a large solid angle which is limited only by the total internal reflectance of the glass support of the printer or photographic plate. The light is attenuated by the density distribution of the developed silver and emerges over a similarly large solid angle which is totally intercepted by the copy material in intimate contact with the surface of the original plate or film (see Fig. A1). This system is remarkably efficient in its utilisation of light, and it enables a tungsten diffuse-light copier to reveal images hidden in densities of around 5.0 with exposure times of a few tens of seconds on slow graphic arts films. It is much more efficient than the distant point-light systems commonly employed for plate copying in astronomy, where the minute solid angle of the illumination severely limits the total flux incident on the original. On the other hand, one might intuitively expect the point-light to offer better resolution than a diffuse light system, but this seems to be qualitatively true only when the density of the original is low, i.e. where the light producing the contact copy image is mainly transmitted rather than mainly scattered through the original emulsion. This property would seem to make the diffuse light copier particularly appropriate for deep astronomical negatives where minimum densities of 1.2–1.5 (i.e. transmissions of 3–6%) are normal.

With originals of low density the diffuse light source exhibits other characteristics which are important in this context. Image grains in unsaturated images tend to be located near the upper surface of a developed photographic emulsion, a property which has been re-examined by Berg (1969). With transillumination, such grains cast very large shadows on to the emulsion surface of a copy film in intimate contact with the original. These shadows are naturally much diluted by the unattenuated light, but by using a high contrast film and with careful control of exposure, the near-surface grains in the original can be recorded on the copy material, but with their effective size much increased. Grains deeper in the emulsion layer are too far away from the copy film to be resolved and do not contribute to the image structure. This effect is particularly important with the fine-grain but quite thick spectroscopic emulsions used in astronomy. Such emulsions are slow and are usually hypersensitised just before use, a process which can give rise to high levels of chemical fog. Unlike the image grains, however, those due to chemical fogging are uniformly distributed throughout the developed layer. Diffuse–light contact printing is a simple way of amplifying the effect of the image grains without at the same time enhancing the non-image noise due to chemical fog.

These distinctive properties of a diffuse-light printer are ideal for exploring the full range of densities on an astronomical emulsion, from the faintest images just detectable above the sky background to the densest features in the blackest parts of the plate.

We chose an Afga–Gevaert SV400 unit which has a useful plate area of 15 × 18 inches (38 × 46 cm). This unit contains twenty-five 25 W tungsten filament opal lamps illuminating a plastic diffuser covered by a thick sheet of glass which forms the work surface. Our copier incorporates a 0–60 s timer and a multi-position switch giving 12 fixed brightness levels which range from about 0.1 to 150 lux on the platten. The unit also incorporates a vacuum pump and gauge and, for consistent results, is run from a stabilisation transformer.

With this simple device two basic methods of contrast manipulation can be undertaken; unsharp masking and photographic amplification. These methods are complementary and can be used in conjunction. However, both methods yield a positive copy as a first derivative and this in turn leads to a third process which can incorporate either masking or amplification, namely additive colour superimposition, which was used to make most of the astronomical colour pictures in this book from black and white negative originals.

All of these methods demand the critical use of photographic materials and careful calibration of the diffuse light copier with limited range of films normally used is essential for predictable and reproducible results. This procedure was outlined by Malin (1977) in his paper on unsharp masking but is here described more fully.

Film and printer calibration

Many of the procedures involved in photographic image manipulation involve making an accurately controlled exposure through an original plate which may have a density anywhere from almost zero to nearly 5.0. It is important to calibrate the response of the copy film to the range of development times employed.

As a reference from which a range of (relative) log exposure values can be obtained, a Kodak No. 1 step wedge is used. It has 21 steps each increasing in density (and thus log E) by 0.15 and is used to contact an image of the steps on to the film to be calibrated. The range of the step wedge can be further increased by covering half of its width with a strip of film (coarse-grained for spectral neutrality) developed to a density of about 2.5. This extends the dynamic range of the step wedge to cover more than 5 log E units (i.e. greater than 100 000:1).

From preliminary tests a development time for the most commonly used film/developer will have been found which produces an acceptable result. On a single sheet of the chosen film, 12 separate but adjoining contact images of the extended step wedge are made at the 12 brightness steps of the contact printer. The exposure time should remain constant, say 5 s for each exposure. The film is carefully processed for the chosen standard time. The results, when plotted in terms of exposure *versus* developed density yield the family of characteristic curves shown in Fig. A2. These curves represent the relative light output of each of the brightness steps of the contact printer and automatically take into account any effects arising as the colour of the lamps alters with brightness.

The next stage is to determine the effect of development time at a fixed exposure level. In this case a piece of the film is given a series of identical contact exposures through the step wedge. The film is processed as a sheet and once every minute a strip

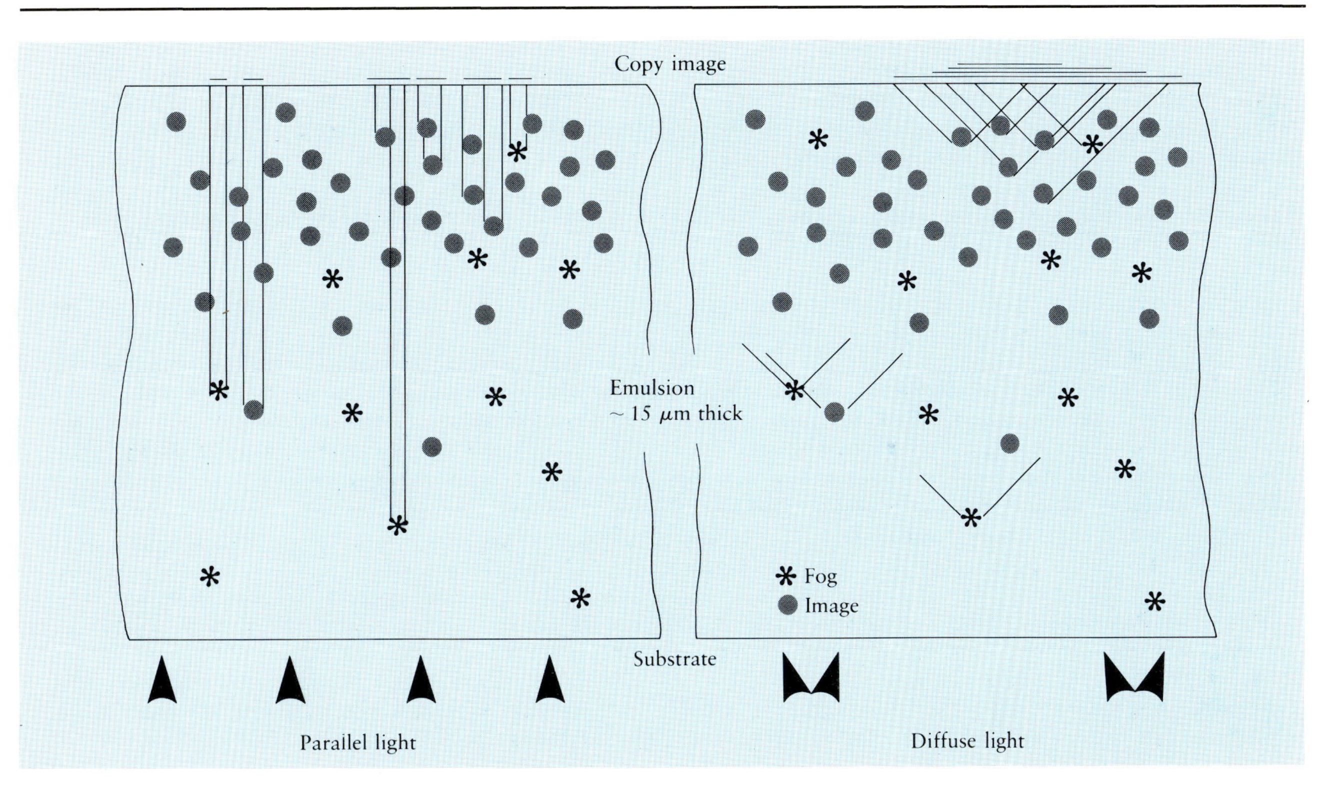

Fig. A1. *(above) The mechanism of photographic amplification. Contact copying by means of parallel illumination (left) from an enlarger or distant point source gives a contact copy where grains throughout the original emulsion are recorded on the copy with equal weight. With diffuse illumination (right) near-surface image grains (solid blobs) appear enlarged on the copy while most fog grains, indistinguishable apart from their distribution, are not detected.*

Fig. A2. *(below) Contact printer calibration curves made on a modified Agfa–Gevaert SV 400 printer. Each curve represents a brightness level on a 12-position step switch for a 5 s exposure. Expressed in terms of density rather than log E, these curves can be used to predict the characteristics of copies made from originals with a very wide range of densities.*

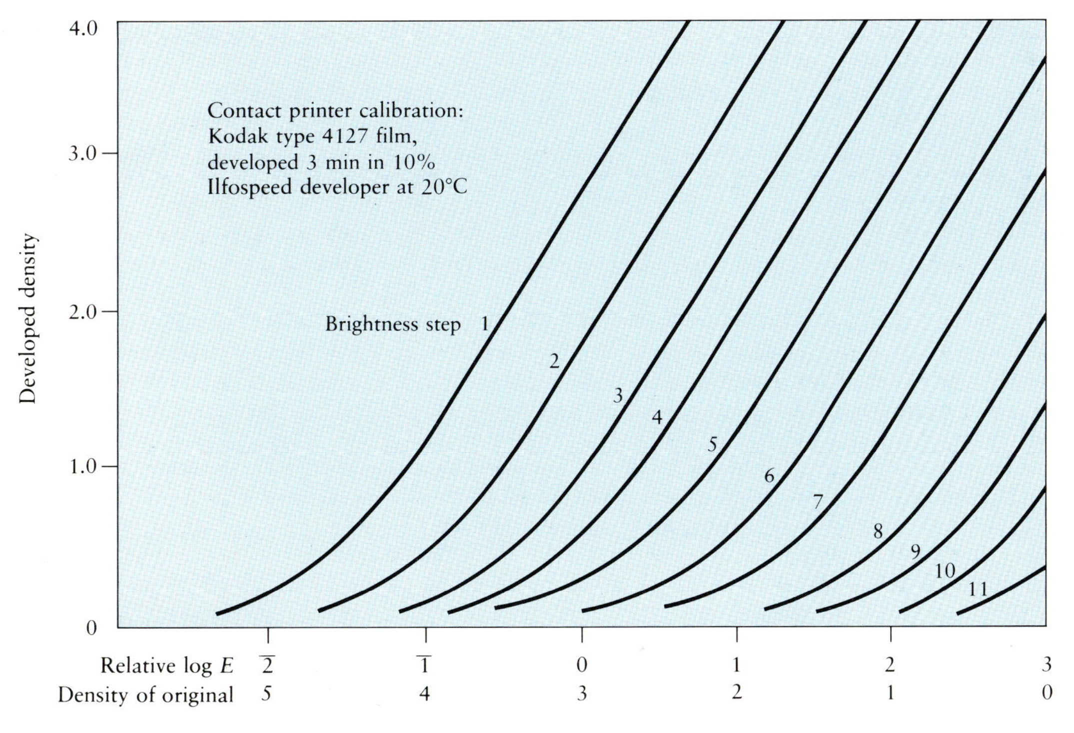

containing a complete but still developing image of the step wedge is snipped off and allowed to fall into an acid stop bath. The resulting strips, ranging in development time from the minimum usable (1 min) to the maximum practicable (about 12 min) are fixed, washed and used to construct the family of development time/density curves shown in Fig A3. From this latter series the development time to reach a given maximum density (D_{max}) amd the time to produce a given contrast (γ) can be derived. Very short development times can result in an image which is yellow-brown in colour due to the incomplete growth of silver grains. Since the films will usually be used with another blue-sensitive bromide paper or copy film this is of no consequence but the density of short development strips should be measured with a blue filter in the densitometer. Non-neutral positives of this kind can not be used for colour separation work.

These fundamentals completed, we are now in a position to make all kinds of copies with predictable and well-defined properties from difficult (i.e. dense and contrasty) originals.

The calibration procedures described above are generally applicable to any film/developer combination likely to be used provided processing is carried out in open trays. Laboratories engaged in work which involves semicommercial-scale reproduction will have access to large-throughput machines for film processing. These are (necessarily) designed for stability rather than flexibility and are not ideal for the rather specialised operations described here.

The developer

When plate copying and the first attempts at image manipulation began in the photolabs at the Anglo-Australian Observatory the need for flexibility led to the use of four or five developers, some of these at a variety of dilutions according to the maker's recommendations. This was unwieldy and in the end meant that no single film/developer combination was thoroughly understood. Since that time it has been found that one developer can be used satisfactorily with the three negative-working and one direct-duplicating films in general use in this laboratory, as well as for developing prints, its intended purpose. This developer, Ilfospeed Print Developer (Ilford Ltd) is available as a concentrated liquid and is conveniently diluted to a 10% solution (100 ml/l, as recommended for prints) or to a 2.5% solution for some copying applications described below.

Kodak Dektol developer, again made three or four times weaker than recommended for prints, is also suitable. It must be emphasised that these formulations are not intended by the manufacturers to be used with graphic arts films, nor are they designed to be used at these extreme dilutions. Nonetheless, experience has shown that consistently reliable results can be achieved with these film/developer combinations provided that the user is aware of the limited capacity and dish life of such low concentrations.

Dish processing

Processing sheet films in open trays is the preferred method where small numbers of copies are to be made. When tray processing under bright red or yellow safelights, a little experience enables one to judge when (for example) some desired maximum has been reached or if density is beginning to appear in the highlights of an image. Development by inspection is a most useful adjunct to the more usual time and temperature methods used for panchromatic films. Panchromatic (and fast ortho) materials are avoided for this reason.

Development is by continuous over-and-over agitation for the full duration of the development time. Of equal importance, and probably more important with short (about 3 min) development times, is the use of an acid stop bath. The agitation here must be vigorous and prolonged, lasting about 30 s before transferring to the fixer.

Unsharp masking

This process involves preparing a blurred positive copy of an original plate and using this as a dodging mask for extracting information from the densest parts of the image. The principle has been used for many years in the printing trade, especially in colour separation (see for example Yule, 1944) but was first used extensively in astronomy by Malin (1977). This paper gives full practical details of the process and describes an earlier version of the calibration procedures, now updated in the previous section. The D76 developer originally used has now been replaced by a 2.5% solution of Ilfospeed concentrate, though the preferred film, Kodak Commercial Film, type 4127, remains unchanged.

This process continues to be a powerful tool for exploring dense images and is capable of yielding important scientific results. The inner shells of NGC 3923 (Chapter 2) are a good example of the way in which hidden features can be revealed by printing a deep plate through an unsharp mask. The first derivative of a plate copied with an unsharp mask is a positive whose contrast and density range can be adjusted to any desired value.

Photographic amplification

This process is simpler to use than unsharp masking and is in a sense complementary, in that here the faintest images are revealed while the denser parts of plates are burned out. The principle of the method was outlined in the previous section on diffuse light

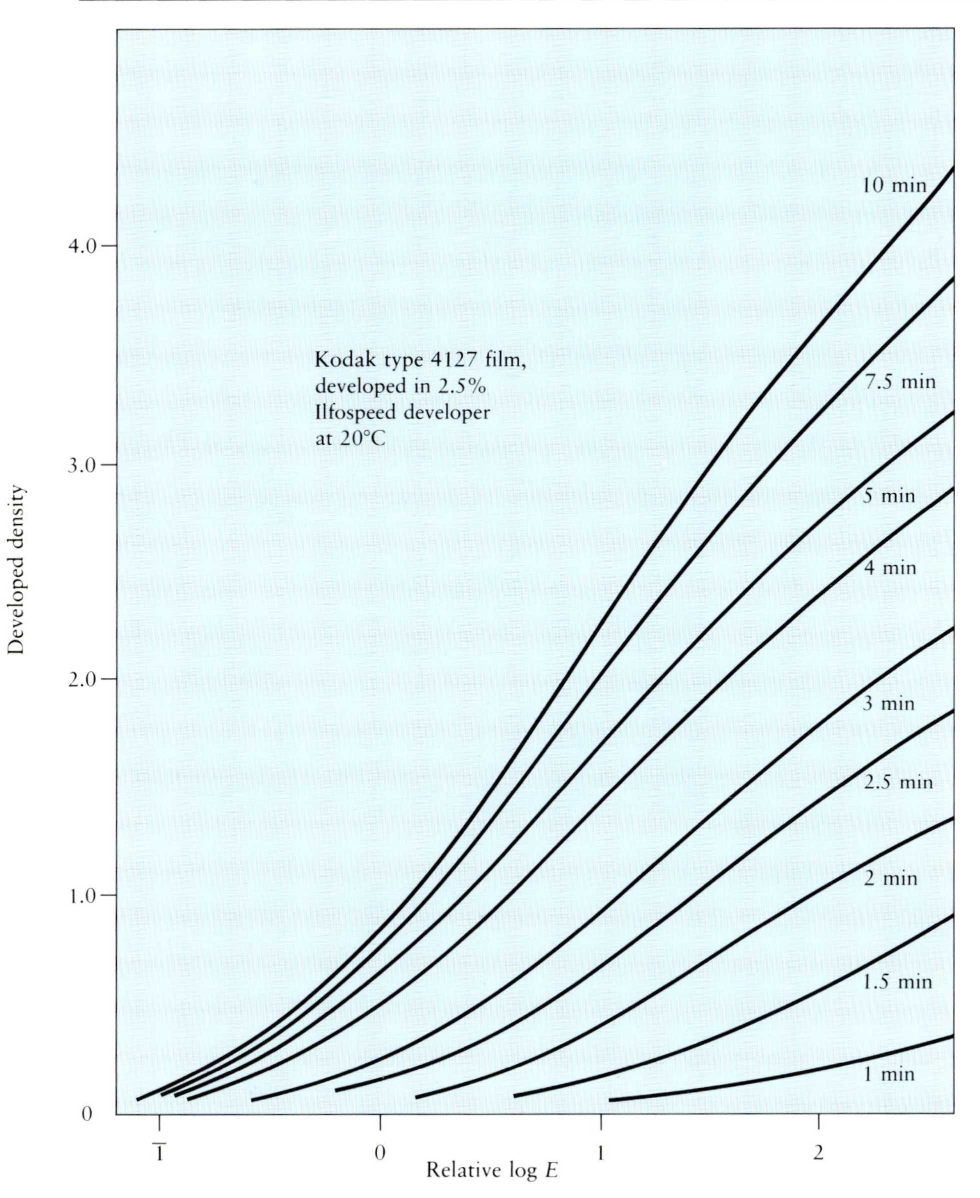

Fig. A3. *At a fixed brightness level, but with variable development time, the change in the characteristic curve of a copy film (Eastman Kodak type 4127) can be shown. The slope (γ) changes from 0.3 to 1.3 as the development time increases from 1 to 12 min. For unsharp masks the slope must always be less than 1.0, corresponding to 4 min development in 2.5% Ilfospeed solution.*

contact copying and a fuller description has been published by Malin (1978*a*). The process has not been changed significantly since publication, although the preferred maximum densities in the processed positives (for monochrome work) are now somewhat lower and the tolerances tighter. For the best results the positives should have maximum densities in the range 0.3–0.4. Where the positives are to be used as colour separations, much higher maximum densities are required, typically 1.2–1.4.

Image superimposition

Images are added together to improve the signal-to-noise ratio when searching for very faint objects. The technique is always used in conjunction with photographically amplified positives – there is little point in combining images which are not already the deepest available. The effect of adding together several images is to reduce the perceived granularity (enhanced by photographic amplification), which is a random and therefore different pattern on each exposure, while reinforcing real images, which repeat from exposure to exposure. An example of image superimposition (or integration printing) is shown on p. 60 (Chapter 2). Combining images is also an essential step in the production of a colour picture from separate monochrome photographs and the solution of this problem was a key development which led to the system used to make most of the colour pictures which appear in this book.

Images are superimposed in the rather simple device illustrated in Fig. A4, which was designed to accept 8 × 10 inch paper or film. It consists of a piece of chip-board (or plywood) about 25 mm thick and 300 mm square. Attached to this surface by means of a good quality (i.e. no free play) piano hinge is a sheet of aluminium 1.5–2 mm thick and 25 mm bigger than the intended paper size. The aluminium must be perfectly flat and when closed the piano hinge is separated from the wooden base by a piece of aluminium the same thickness as the lid. A further strip of aluminium (shown attached with four screws) is to assist location of the unexposed paper in the dark. The lid is weighted with a strip of angle-iron running along its free edge which also acts as a lifting handle. If the lid is really flat, the frame is light – tight enough to permit full white room lighting when loaded with film or paper. This device can be constructed in any home workshop for less than £5.

The superimposition frame is always used with a projected image from an enlarger. The area of interest is aligned, framed and focused on a sheet of white paper attached to the working surface beneath the hinged lid. This surface is made slightly tacky with a non-drying spray print adhesive. The hinged lid is closed and a piece of unexposed resin-coated (i.e. dimensionally stable) bromide paper is lightly affixed to the upper surface which is also slightly tacky. An exposure is made to produce a print of good contrast, which after processing and drying is replaced on the lid as shown in the illustration. With the reference print in place, a sheet of unexposed paper or film is located beneath the lid which is then firmly closed. The reference print is aligned with the projected image by moving the whole frame around on the enlarger baseboard; the 'fine adjustment' is made by a gently tapping action with the finger-tips. When properly aligned, the image on the reference print and that projected from the enlarger cancel completely and quite suddenly. This feature leads to extremely good and reproducible superimposition. The hinged lid is carefully lifted and a timed exposure is made on the unexposed material in the normal way. Subsequent images are aligned and exposed in a similar fashion. In practice this arrangement is very easy to use; the total 'assembly' time for a three-image superimposition is less than 10 min. Exposure times for combining high-contrast positives must be found by trial and error but with due weighting given to slightly denser positives so that the contribution to the sky background density is the same from each image. Exposure for colour pictures is discussed in Appendix II.

Apart from its value in image superimposition, this procedure is an excellent way of detecting small vibrations in the enlarger due to air conditioning drafts or heavy machinery nearby. It also detects scale changes due to small temperature shifts and (by turning the negative and frame through 180° between exposures) any out-of-square components in the enlarger system.

Fig. A4. *A simple superimposition frame, shown with a reference print attached to the hinged lid. All surfaces beneath the lid are blackened to eliminate light leaks. A device of this kind opens up many new opportunities for picture-making.*

Appendix II: Additive colour photography

The original plates

The rationale which led us to the use of a colour separation process for colour pictures of astronomical objects is dealt with in Chapter 2. Here we are concerned with the practical details of Maxwell's three-colour additive process brought up-to-date and used with the spectroscopic plates specially designed for astronomy.

To use as much archival material as possible and to ensure that plates exposed for colour pictures could be employed for other purposes we used standard emulsion/filter combinations corresponding to the well-known B, V and R photographic photometry colours. This also has the advantage that colour separation plates fit easily into the routine of a night's observing. Passbands were defined by the following Kodak emulsion types and Schott filters:

Blue – IIaO emulsion with Schott GG385 or 395 filter (385 or 395–500 nm)

Green – IIaD emulsion with Schott GG475 or 495 filter (475 or 495–630 nm)

Red – 098–04 emulsion with Schott RG610 or 630 filter (610 or 630–690 nm)

With these combinations, a passband is defined at the blueward end by the filter and redwards by the spectral sensitivity of the emulsion. The emulsions are essentially non-colour-sensitised (IIaO), orthochromatic (IIaD) and extended-red panchromatic (098–04). The Schott glass filters used at the AAT are 2 mm thick while those used on the UK Schmidt are 3 mm thick, thus the blue-end cut-on is slightly different for each telescope.

The matching of the positives and final colour balancing at the printing stage was greatly simplified by the inclusion of a calibrated grey scale on each AAT exposure. The 16-spot pattern, covering an input brightness range of about 1000:1 (three log *E* units) was imaged on to the plate by an optical projector similar to that described by Schoening (1976). The grey scale appears in a corner of the plate in a region protected from the telescope image. Filtered to a colour temperature of 5500K (mean noon sunlight) the step wedge image passes through both telescope filter and shutter before reaching the plate. The calibration and telescope exposures are therefore identical in all respects.

Careful sensitometric testing of each batch of plates after hypersensitisation and exposure in a nitrogen atmosphere, together with a stabilised power supply for the projection calibrator permits telescope exposures to be predicted so that all three exposures (B, V and R) are identical with respect to the developed density in one or more of the calibrator steps.

The exposure times were chosen to give sky background densities of 0.6–0.8 (ANSI diffuse) above chemical fog. Plates exposed to this density yield the maximum output signal-to-noise for emsulsions of this type (Latham, 1974; Hoag *et al.*, 1978). On the 1.2 m UK Schmidt Telescope with a focal ratio of f/2.5, exposure times were 40–60 min for each colour on unhypersensitised emulsions. At the f/3.3 prime focus of the 3.9 m Anglo-Australian Telescope the same sky-limited condition was reached in 30–40 min after hypersensitising. AAT plates are hypersensitised by baking at 65 °C in a flow of nitrogen followed by a few hours in hydrogen at 20 °C. The hypersensitised plates were exposed in a nitrogen atmosphere to prevent speed deterioration due to ambient moisture

(Malin, 1978*b*). At both establishments plates were developed with a Palomar-type (Miller, 1971) tray rocker in D19 for 5 min at 20 °C.

Subsequent combination of the colour separation positives would have been simplified if one emulsion type had been used for all three exposures (i.e. IIaO, IIaD and IIaF or 103aO, D and F). The slope of the characteristic curves (contrast) would have been very similar (although probably not identical – see Farnell, 1959) and matched positives could be made with equivalent exposure and equal development for all three colours. Unfortunately emulsion type 103a is not used at either UKST or the AAT but IIaO and IIaD are available. Experiments with IIaF showed it to be slow and in our hands incapable of significant improvement by any of our standard hypersensitising procedures. We therefore chose to use Kodak's 098–04 (an improved version of 103aF) for the red exposures. This material has a higher contrast ($\gamma = 1.9$) than the IIa types ($\gamma = 1.5$) and is somewhat grainier but it proved possible to match the curve shapes of all three exposures during the copying stage. The characteristic curves of the originals can however be made much more compatible by using a development time longer than that recommended for the IIaO and IIaD plates.

We found that an extension of the development time for the IIa's to 8 min gave a much closer match for the three emulsions in terms of both contrast and granularity. Some reduction in the B and V exposure times is an advantage of this approach.

In our first experiments with three-colour separation negatives this equivalence of the three exposures with respect to the step wedge was considered essential. Experience has shown however that the copying stage is sufficiently flexible to cope with wide variations in contrast, fog and density in the original plates. It is important to note that in sky-limited exposures, with a minimum density on the original of about 0.6 above chemical fog all the image information is on the straight-line part of the characteristic curves – the toe region can be ignored, which greatly simplifies contrast adjustment.

Positive copies and contrast adjustment

In general, making the positives involves reducing the contrast of the originals and matching their characteristic curves so that all three are as parallel as possible. This usually involves adjusting copy exposure and development to bring the curves of the plates on similar emulsions (IIaO and IIaD) closer together. Depending on the density range to be included in the final print, it is often possible to adjust the curve of the higher contrast red plate (098–04) in the same way by reducing the development time of its copy positive. Where this cannot be done, a pre-fogging technique may be used to alter the shape of the toe region of the positive emulsion. Before the contact exposure to the original the copy film is given a pre-exposure (usually under the enlarger) to produce a uniform fog of density 0.05–0.2. This has the effect of lowering the contrast in the highlights of the developed positive without loss of information near the sky background. The effect of these procedures on a set of plates with widely different original exposures is shown in Fig. A5.

The flexibility of these copying methods can be further enhanced by the incorporation of more advanced methods of image manipulation based on the positive-working contact techniques referred to on p. 186. The colour photographs of the Orion Nebula (p. 108) were made from dense plates by means of unsharp masks made individually for each of the three original plates and photographic amplification was used to enhance the faint whisps of the Vela supernova remnant on p. 105.

Superimposition

At Maxwell's demonstration of the first colour photograph in 1861 he superimposed red, green and blue images by juggling three 'magic lanterns' so that their projected images were in register on a reflecting screen. Because the optical axes of three projectors cannot be made to coincide without an elaborate arrangement of half-silvered mirrors it is unlikely that Maxwell achieved perfect registration.

Much better results can be obtained by using a single projector (i.e. an enlarger) and superimposing the coloured positive images one after the other on to a material which is capable of recording a positive, coloured image. This option was not, of course, available to Maxwell and only in recent years have positive-working colour materials with simple processing been generally available.

The images are superimposed in register using the simple hinged-lid device illustrated on p. 188. With one of the positive colour separations in the enlarger, the image to be made into a colour picture is chosen and the frame aligned with it. A piece of RC bromide paper is attached to the lid and an exposure is made to produce a reference negative print. When this paper is processed, dried and refixed to the lid (position not critical), a piece of positive-working colour paper can be inserted into the superimposition frame and the light-tight, hinged cover carrying the reference print is closed. The reference print, firmly attached to the lid of the frame, is now aligned with the projected image of one of the colour separations, a process which can be done more easily without colour filters in the enlarger. With perfect registration (i.e. cancellation of the negative print and positive projected image), the appropriate colour filter is inserted, the lid opened and an exposure in (say) blue light is made through the blue positive on to the colour paper. The positives

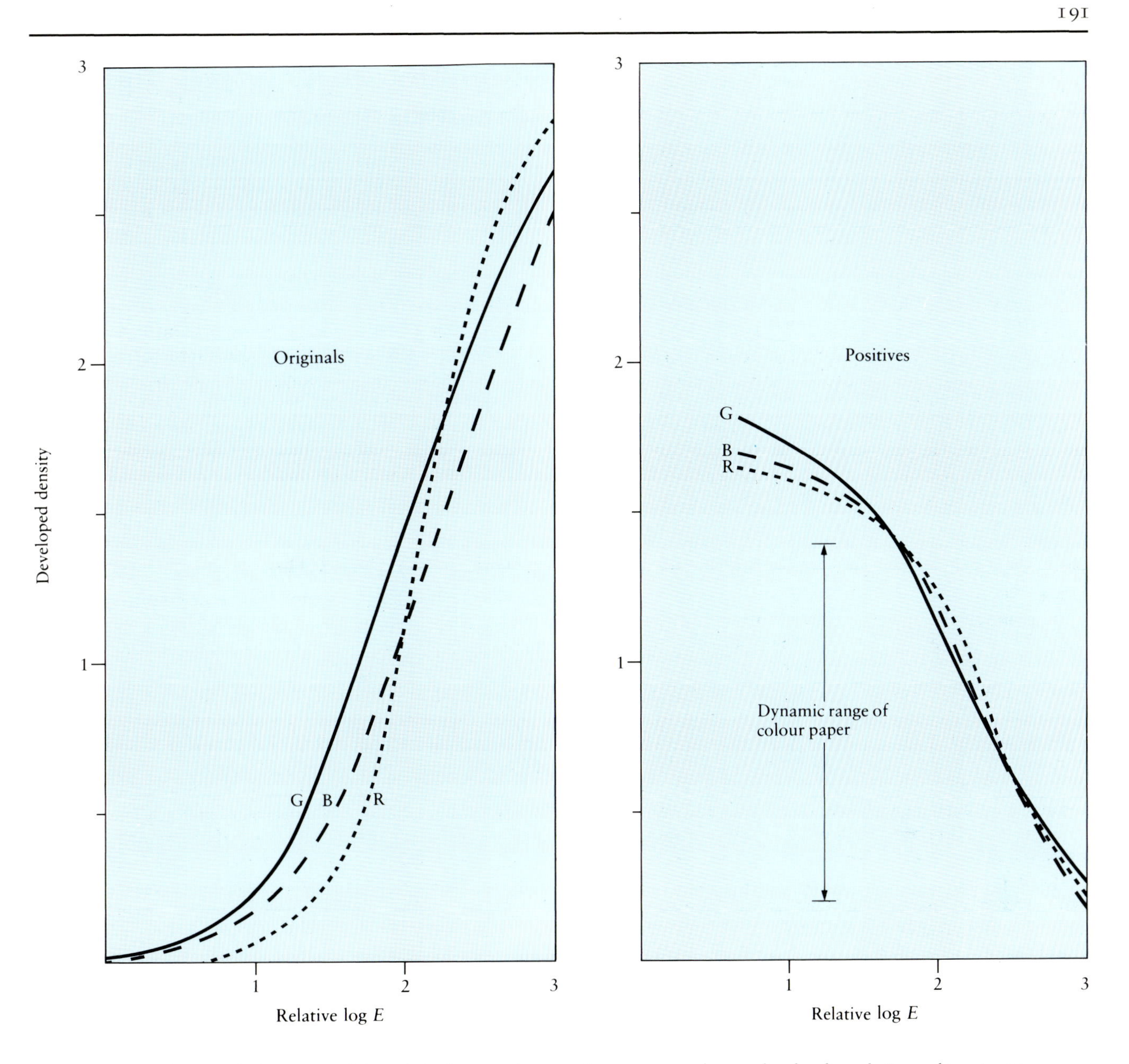

Fig. A5. *The characteristic curves of the original monochrome plates are not always closely aligned. By making positive copies, the curves can be brought nearer together with a reduction in contrast. The toe region of the curves for the originals is usually ignored because the minimum density on deep plates brings all the information on to the straight line portion.*

made from originals taken in green and red light are likewise aligned and exposed separately through green and red filters.

Initial experiments with a photographic grey scale in the enlarger had shown the relative blue/green/red exposure times required to reproduce a neutral grey on the colour-sensitive paper. With a conventional 300 W opal enlarger lamp operating at 2700K and Wratten filters 47B (blue), 58 (green) and 25 (red) the exposure times were almost exactly in the ratio 3:2:1. The grey scale also revealed that the colour paper then in use had a tonal range equivalent to 1.2 density units in the positive copy with a normal condenser enlarger. The positive copies were therefore made with a tonal range, ΔD of 1.2. With these basic parameters established, minor colour correction, to compensate for small departures from the ideal in the positives are easily made by altering the relative exposure times of the three colours. The small batch-to-batch variations in paper colour balance are likewise accommodated. The grey scale on the originals, transferred through the copying processes, is a sensitive colour-balance reference. Representative colour balance is achieved by ensuring that this grey scale remains neutral in the final colour composite.

All the three-colour photographs which appear in this book were made using Cibachrome paper as the positive-working colour 'receiver' for the separate monochromes. Cibachrome paper has two advantages which recommend it for this kind of work. It is relatively slow, which means that a bright image can be used on the superimposition frame for setting up and adjustment without altering the enlarger lens aperture. In addition the three-solution processing cycle is rapid (about 10 min), requires no special equipment and is not critically temperature-dependent. There is however absolutely no reason why other positive-working processes should not be used although the effective dynamic range of each should be determined before making the separation positives.

In our hands this process has proved to be flexible and versatile, incorporating as it does the control necessary to make colour pictures of astronomical objects. The processes in themselves are simple but combining all of them into a workable colour system requires some skill, patience and experience, none of which can be learned from books. This appendix will have served its purpose if it contains sufficient information to encourage others to try the techniques.

Summary of materials and processes

1. For unsharp masks and low-contrast copies ($\gamma = 0.5$–1.0): Kodak Commercial film type 4127, developed in a 2.5% solution of Ilfospeed developer, 1–5 min, 20 °C.
2. For colour separations of moderate contrast ($\gamma = 0.8$–1.5): Kodak Commercial film type 4127, developed in 10% Ilfospeed, 1–5 min, 20 °C.
3. For higher contrast ($\gamma = 2.5$–4.0): Ilford Line film, developed in 10% Ilfospeed, 20 °C.
4. For photographic amplification and extreme contrast ($\gamma = 5$–7): Agfa–Gevaert F071p ortho film *or* Kodak Reproduction film type 2566, developed in 10% Ilfospeed, 2–4 min, 20 °C.
5. For direct reversal, variable contrast ($\gamma = 0.8$–10): Dupont Direct Positive film Cronalar SD 4 *or* Kodak Precision Line film LPD 7, developed 2–4 min in 10% Ilfospeed. Contrast is varied by a uniform pre-flashing before the image exposure. See Malin (1982).

Bibliography

In the text we have given references at places where we think readers might want to follow up what we have described. Generally the descriptive astronomy or the well-known history of astronomy is not referenced, but original historical works which are not well known, and works useful to astronomers on photography, colour, theory of vision, etc., are listed in this bibliography. The bibliography on the redness of Sirius is an attempt to be complete.

Abbot, C.G. and Fowle, F.E. (1913) Volcanoes & climate. *Ann. Astrophys. Obs. Smithsonian Inst.*, 3, 211.

Allen, D.A., Carter, D. and Malin, D.F. (1982) The nature of the shells of NGC 1344. *Nature*, **295**, 126–8.

Alt, E., Brodkorb, E. and Rusche, J. (1974) More about indirect color. *Sky and Tel.*, **48** (5), 333–8.

Arago, F. (1855) Plate X. In *Popular astronomy*, vol. 1, Longman, Brown, Green and Longman, London.

Araya, G., Blanco, V.M. and Smith, M.G. (1972) A new extension of the Helix Nebula. *Publ. Astr. Soc. Pacific*, 84, 70.

Babcock, T.A. (1976) A review of methods and mechanisms of hypersensitization. *AAS Photo-Bull.*, No 13, 3–8.

Babcock, T.A., Sewell, M.H., Lewis, W.C. and James, T.H. (1974) Hypersensitization of spectroscopic films and plates using hydrogen gas. *Astron. J.*, 79, 1479–87.

Barker, T. (1760) **51**, *Philosophical Transactions*, Remarks on the mutations of the stars. 498.

Barnard, E.E. (1890*a*) On the photographs of the Milky Way made at the Lick Observatory in 1889. *Publ. Astr. Soc. Pacific*, 2, 240–4.

Barnard, E.E. (1890*b*) On some celestial photographs made with a large portrait lens at the Lick Observatory. *Mon. Not. R. astr. Soc.*, 50, 310–14.

Barnard, E.E. (1919) On the dark markings of the sky with a catalogue of 182 such objects. *Astrophys. J.*, 49, 1–23.

Barnard, E.E. (1927) *A photographic atlas of selected regions of the Milky Way*. Washington: Carnegie Institute.

Baum, R. (1979) Historical sighting of the craters of Mercury. *J. Ass. Lunar & Planetary Observers*, 28, 17.

Berg, W.F. (1969) The photographic emulsion layer as a three-dimensional recording medium. *Appl. Optics*, 8, 2407–16.

Bok, B.J. and Reilly, E.F. (1947) Small dark nebulae. *Astrophys. J.*, 105, 255–7.

Boll, F. (1917) Astronomical observations in antiquity. *Neue Jahrbucher für das Klassischen Alterum*. Leipzig, p. 20.

Bouman, M.A. (1961) Quantum theory in vision. In *Sensory communication*, ed. W.A. Rosenblith, pp. 377–402. New York: Wiley.

Bowen, I.S. and Clark, L.T. (1940) Hypersensitization and reciprocity failure of photographic plates. *J. Opt. Soc. Am.*, 30, 508.

Chambers, G.F. (1877) *The starry heavens*. Oxford University Press.

Cohen, H.L. and Oliver, J.P. (1981) Star colors: an astronomical myth? *Sky & Telescope*, **61**, 104.

Common, A.A. (1883) Note on a photograph of the Great Nebula in Orion and some new stars near Θ Orionis. *Mon. Not. R. astr. Soc.*, 43, 255–7.

Cros, C. (1869) Solution du problème de la photographiè des couleurs. *Les Mondes*, Feb. 25.

D'Antona, F. and Mazzitelli, I. (1978) Constraints on the corona model for Sirius B. *Nature*, **275**, 726.

Davies, E.R. (1936) The Kodachrome process of 16 mm colour kinematography. *Phot. J.*, 76, 248.

de Vaucouleurs, G., de Vaucouleurs, A. and Freeman, K.L. (1968) The Magellanic Barred Spiral Galaxy NGC 4027. *Mon. Not. R. astr. Soc.*, 139, 425.

Dittrich, E. (1927) Where the epithet red for Sirius originates. *Astr. Nachrichten*, **231**, No. 5542.

Dopita, M., *et al.* (1980) N70; a mass-loss bubble within a massive collapsing HI cloud. *Astrophys. J.*, **250**, 103.

Draper, J.W. (1840) Dageurrotypes of the Moon. *Phil. Mag.*, 17, 217–25.

Ducos de Hauron, L. (1869) *Les couleurs en photographie: solution du problème*. Paris: A. Marion.

Dufour, R.I. and Goodding, R.A. (1976) Color composite photograph of the Magellanic Clouds. *Proc. S. W. Reg. Conf. Astron. Astrophys.*, (July 12, Lubbock, Texas) pp. 71–4.

Dufour, R.I. and Martins, D.H. (1976) Reconstructing color images of astronomical objects using black and white spectroscopic emulsions. *J. Appl. Phot. Eng.*, **2**, 93–4.
Eastman Kodak (1973) *Publication F5: Emulsion characteristics of Plus-X film.*
Eddington, A.S. (1911) Star. In *Encyclopaedia Brittanica*, **25**, 788.
Edwards, A. (1876) A new method of printing from plates of gelatin. *English Patent* 1362.
Evans, R.M. (1961*a*) Some notes on Maxwell's color photograph. *J. phot. Sci.*, **9**, 243–6.
Evans, R.M. (1961*b*) Maxwell's color photograph. *Sci. Am.*, **205** (5), 118–28.
Fang Li-Zhi (1961) A brief on astrophysics in China today. *Chinese Astronomy & Astrophysics*, **5**, 1.
Farnell, G.C. (1959) Relationship between sensitometric and optical properties of photographic emulsion layers with particular reference to the wavelength variation of sensitivity. *J. phot. Sci.*, **7**, 83–92. (*see also J. phot. Sci.*, **2**, 145–9 (1954).
Foucault, M.M. and Fizeau, H.L. (1845) *Popular Astronomy*, **XII**, 1.
Gascoigne, S.C.B. (1966) Colour magnitude diagrams in the Magellanic Clouds. *Mon. Not. R. astr. Soc.*, **134**, 59.
Grassman, H.G. (1853) Zur Theorie der Farbenmischung. *Poggendorffs Ann. Phys.* **89**, 69.
Grassman, H.G. (1854) Theory of compound colour. *Phil. Mag.*, ser. 4, **4**, 254–64.
Gill, D. (1882) Photographs of the Great Comet. *Mon. Not. R. astr. Soc.*, **43**, 53.
Guild, J. (1931) The colourimetric properties of the spectrum. *Phil. Trans. Royal Soc. London*, **A230**, 149–87.
Gundel, W. (1927) Sirius. In *Real Encyclopaedie der Classichen Altertumswissenchaft*. Stuttgart, Fünfter–Halband.
Hamilton, J.F. (1977) Reciprocity failure and the intermittency effect. In *The theory of the photographic process*, 4th edn, ed. T.H. James, pp. 133–145. New York: Macmillan.
Hecht, S., Shlaer, S. and Pineune, M.H. (1942) Energy, quanta and vision. *J. Gen. Physiol.*, **25**, 819–40.
Helmholtz, H., von (1860) *Handbuch der Physiologischen Optik*. Hamburg: Voss.
Henbest, N. and Marten, M. (1983) *The New Astronomy*. Cambridge: Cambridge University Press.
Herschel, J. (1839) Brief des Baronets. *Astr. Nachrichten*, No. **372**, 187.
Herschel, W. (1785) On the construction of the heavens. *Phil. Trans*, **75**, 213–66.
Herschel, W. (1791) On nebulous stars, properly so called. *Phil. Trans.*, **81**, 71–88.
Herschel, W. (1814) Astronomical observations relating to the sidereal part of the heavens. *Phil. Trans.*, **104**, 248–84.
Hoag, A.A. (1960) Low-temperature modification of reciprocity failure in astronomical photography. *Publ. Astr. Soc. Pacific*, **72**, 352–3.
Hoag, A.A. (1961) Cooled emulsion experiments. *Publ. Astr. Soc. Pacific*, **73**, 301–8.
Hoag, A.A., Furenlid, I. and Schoening, W.E. (1978) Think S/N! *A A S Photo Bull*, No. **19**, 3–6.
Hoffleit, D. (1966) *Catalogue of bright stars*. New Haven, Yale U.P.
Hoskin, M.A. (1963) *William Herschel and the construction of the heavens*. London: Oldbourne Press.
Hoyt, W.G. (1981) T.J.J. See and Mercurian craters. *J. History Astr.*, **12**, 139–142.
Humboldt, G. (1845) *Cosmos*, **III** (I), p. 111.
Hunt, R.G.W. (1952) Light and dark adaptation and the perception of colour. *J. Opt. Soc. Am.*, **42** (3), 190–9.
Hunt, R.W.G. (1975) *The reproduction of colour*, 3rd edn. Kings Langley: Fountain Press.
Hurter, F. and Driffield, V.C. (1890) Photochemical investigations and a new method of determination of the sensitiveness of photographic plates. *J. Soc. Chem. Ind. (London)*, **9**, 455.
James, T.H. (1972) Effects of moisture and oxygen on inherent and spectral sensitivity. *J. phot. Sci.*, **20**, 182–6.
Joly, J. (1894) On colour photography. *Brit. J. Phot.*, **41**, 457–9.
Klein, H.J. (1867) Ueber den Farbenwechsel einiger Fixsterne. *Astr. Nachrichten*, **70**, 105.
Klein, H.J. (1876) Ueber den Periodischen Farbenwechsel von α UMa. *Astr. Nachrichten*, **88**, 363.
Kohler, R.J., Howell, H.K. (1963) Photographic image enhancement by the superimposition of multiple images. *Photogr. Sci. Eng.*, **7**, 241.
Kopal, Z. (1959) *Close binary systems*. London: Chapman & Hall.
Lankford, J. (1980) A note on T.J.J. See's observations of craters on Mercury. *J. History Astr.*, **11**, 129–32.
Latham, D.W. (1974) Detective performance of spectroscopic plates. In *Methods of experimental physics*, vol. 12. *(Astrophysics, Part A, Optical and IR)*, ed. N. Carleton. London: Academic Press.
Lauterborn, D. (1970) In *Mass-loss and evolution in close binaries (IAU Coll. 6)*, ed. K. Gyldenkerne & R.M. West, p. 190. Copenhagen: Copenhagen University Press.
Lauterborn, D. (1971) Evolution with mass exchange of Case C for a binary system of total mass 7 solar masses. *Astr. & Astrophys.*, **7**, 150–9.
Lewis, T. (1906) Measures of Struve double stars. *Mem. R. astr. Soc.*, **56**, 1.
Lindegren, L. and Dravins, D. (1978) Holography at the telescope: an interferometric method for recording stellar spectra in thick photographic emulsions. *Astron. Astrophys.*, **67**, 241–55.
Lindenblad, I.W. (1975) On K.D. Rakos' measurements of Sirius B. *Astr. & Astrophys.*, **41**, 111–12.
Lippmann, G. (1891) La photographie des couleurs. *Compt. Rend. Acad. Sci. Paris*, **112**, 274.
Lippmann, G. (1894) La théorie de la photographie des couleurs. *J. de Phys.*, 3:e ser., **3**, 97.
Lynn, W.T. (1887) The alleged ancient red colour of Sirius. *Observatory*, **10**, 104–5.
MacAdam, D.L. (1970) *Sources of color science*. Cambridge: MIT Press.
Malin, D.F. (1977) Unsharp masking. *A A S Photo Bull*, No. 16, 10–13.
Malin, D.F. (1978*a*) Photographic amplification of faint astronomical images. *Nature*, **276**, 591–3.
Malin, D.F. (1978*b*) The effect of environment on the sensitivity of hydrogen hypersensitized spectroscopic plates. In *Modern techniques in astronomical photography, ESO Conference Proc., Geneva, May 1978*, p. 107–112. Geneva: ESO.
Malin, D.F. (1979) A jet associated with M89. *Nature*, **277**, 279–80.
Malin, D.F. (1980) Southern skies explored. *Astronomy* (USA), **8**, 6–14 (Feb).
Malin, D.F. (1982) Photographic enhancement of direct photographic images. *A A S Photo Bull.* **27**, 4–9.
Malin, D.F. (1982*a*) Direct photographic image enhancement in astronomy. *J. Phot. Sci.* **29**, 199–205.

Malin, D.F. (1982*b*) Photographic image intensification and reduction: a unified optical approach. *J. Phot. Sci.*, **30**, 87–94.

Malin, D.F. and Carter, D. (1980) Giant shells around elliptical galaxies. *Nature*, **285**, 643–5.

Maran, S.P. (1975) Red, white and mysterious. *Natural History*, Aug–Sep.

Marchant, J.C. and Millikan, A.G. (1965) Photographic detection of faint stellar objects. *J. Opt. Soc. Am.*, **55**, 907.

Maxwell, J.C. (1860) On the theory of compound colours and the relations of colours of the spectrum. *Proc. Royal. Soc.*, **10**, 404, 484.

Maxwell, J.C. (1861) On the theory of the three primary colours. *Brit. J. Phot.*, **8**, 270.

Miller, J.S. (1976) The structure of emission nebulas. *Sci. Am.*, **231**, 34–43.

Miller, W.C. (1959) First color portraits of the heavens. *Nat. Geo. Mag.*, (May) 670–9.

Miller, W.C. (1961) Celestial wonders the eye has never seen. *Ansconian* (GAF House Magazine), **3**, 1–9.

Miller, W.C. (1962) Color photography in astronomy. *Publ. Astr. Soc. Pacific.*, **74**, 457–73.

Miller, W.C. (1971) Photographic processing for maximum uniformity and efficiency. *A A S Photo Bulletin*, No. 4 (1971, No. 2), 3–18.

Minnaert, M. (1954) *Light and colour in the open air*. New York: Dover Publications.

Mitton, S. (1977) A new look at southern skies. *Illust. London News*, **265** (6953), 39–42.

Murdin, P.G., Allen, D.A. and Malin, D.F. (1979) *Catalogue of the Universe*. Cambridge University Press.

Neckel, T. (1978) Observations of stars in the H II regions NGC 6334 and NGC 6357. *Astr. & Astrophys.*, **69**, 51–6.

Nelson, C.N. (1966) The theory of tone reproduction. In *The theory of the photographic process*, 3rd edn, ed. T.H. James, pp. 495–7. New York: Macmillan.

Newcomb, S. (1902) *The stars*. London: Murray.

Osthoff, H. (1927) Concerning the colour of Sirius in Antiquity. *Astr. Nachrichten*, NO. 5495.

Paczinski, B. (1970) In *Mass-loss and evolution in close binaries (IAU Coll. 6)* ed. K. Gyldenkerne & R.M. West, p. 192. Copenhagen: Copenhagen University Press.

Racine, R. (1968) Stars in reflection nebulae *Astron. J.*, **73**, 233.

Rakos, K.D. (1974) Photoelectric measurements of Sirius B. *Astr. & Astrophys*, **34**, 157.

Rhim, K. (1974) Color portraits of deep-sky objects. *Sky and Tel.*, **48** (2), 120–4.

Richter, N. (1975) Optimisation of information in astronomical photography. *Vistas in Astronomy*, **19**, 215–23.

Roark, T.P. and Greenberg, J.M. (1967) Models of reflection nebulae. In *Interstellar grains*. NASA: Washington D.C.

Sandage, A. and Miller, W.C. (1966) A search for a cluster of galaxies associated with 3C 48 using the Kodak special plate 087–01. *Astrophys. J.*, **144**, 1238–40.

Schiaparelli, G.V. (1896) Rubra Canicula. *Atti della Academia di Scienze degli Agiati de Rovereto*, Ser III, Vol. II, Fasc. II and Vol. III, Fasc. I & II.

Schoening, W.E. (1976) Final design of the Kitt Peak projection spot sensitometer. *A A S Photo Bulletin*, No. 11, (1970, No. 1), 8–10.

See, T.J.J. (1892) History of the colour of Sirius. *Astr. & Astrophys.*, **11**, 269–274, 372, 457, 550.

See, T.J.J. (1898) The Fundamental law of temperature for gaseous bodies. *Astronomical J.*, **19**, 181–185.

See, T.J.J. (1927) Historical researches indicating a change in the colour of Sirius, between the epochs of Ptolemy, 138, and of al Sûfi, 980 AD. *Astr. Nachrichten*, **229**, 245–72.

Sim, M.E., Hawarden, T.G. and Cannon, R.D. (1976) Photographic plate sensitizing techniques at UKSTU. *A A S Photo Bulletin*, No. 11, (1976, No. 1), 3–6.

Smith, A.G. and Hoag, A.A. (1979) Advances in astronomical photography at low light levels. *Ann. Rev. Astron. Astrophys.*, **17**, 43–71.

Smith, A.G. and Schrader, H.W. (1979) Balanced hypersensitization of a fast color reversal film. *A A S Photo Bull*, No. 21, 9–14.

Smyth, W.H. (1844) *Cycle of celestial objects*, London: J. Parker.

Smyth, W.H. (1864) *Sidereal chromatics or colours of double-stars*, London: J. Bowyer Nichols.

Stone, E.J. (1884) Presentation of the Society's Gold Medal to Mr Common. *Mon. Not. R. astr. Soc.*, **44**, 221.

Stone, R.P. (1979). The Crossley Reflector. *Sky & Telescope*, **58**, 307, 396.

Struve, F.G.W. (1827) *Catalogus novus stellarum duplicium et multiplicium*, Dorpat, Estonia.

Struve, O., Elbey, C.T. and Roach, F.E. (1936) Reflection nebulae. *Astrophys. J.*, **84**, 219.

Sutton, T. (1861) *Photographic Notes*, **6**, 125, 169–70.

Temple, R. (1975) Letter of 13 August 1975.

Tull, A.G. (1963) Tanning development and its application to dye transfer images. (An extensive bibliography on dye transfer processes.) *J. Phot. Sci.*, **11**, 1.

Turner, H.H. (1899) On the fundamental law of gaseous reputation. *Observatory*, **22**, 292.

van den Bergh, S. (1966) Catalogue of reflection nebulae. *Astron. J.*, **71**, 990–8.

van den Bergh, S. (1970) Extragalactic distance scale. *Nature*, **225**, 503.

van der Post, L. (1963) *The seed and the sower*. London: Hogarth Press.

Vogel, H. (1873*a*) Uber die Lichtempfindlichkeit des Bromsilbers für die sogenannten chemisch unwirksamen Farben. *Berichte der D C G*, **6**, 1302.

Vogel, H. (1873*b*) On the sensitiveness of bromide of silver to the so-called chemically inactive colours. *Photo. News*, **17**, 589.

Walker, M.F. (1967) A two-color dye-transfer photograph of M33. *Publ. Astr. Soc. Pacific*, **79**, 119–24.

Walker, M.F., Blanco, J.M. and Kuukel, W.E. (1969) Two-color composite photographs of the Magellanic Clouds. *Astron. J.*, **74**, 44–6.

Walls, G.L. (1956) The G. Palmer Story. *J. Hist. Med.*, **11**, 66–96.

Webb, J.H. (1935) The effect of temperature upon reciprocity law failure in photographic exposures. *J. Opt. Soc. Amer.*, **25**, 4.

Webb, T.W. (1851) *Celestial objects for common telescopes*. New York: Dover.

Weitzman, D.O. and Kinney, J.A.S. (1969) Effect of stimulus size, duration and retinal location upon the appearance of color. *J. Opt. Soc. Amer.*, **59**, 640.

Wright, W.D. (1928) A redetermination of the trichromatic coefficients of the spectral colours. *Trans. Opt. Soc. London*, **30**, 141–64.

Young, T. (1802) On the theory of light and colours. *Phil. Trans. Royal Soc. Lond.*, **92**, 20–71.

Yule, J.A.C. (1944) Unsharp masks. *Phot. J.*, **84**, 321–7.

Index

QB
816
.M34
1984

Malin, David.
Colours of the stars / by David Malin and Paul Murdin. -- Cambridge ; New York : Cambridge University Press, 1984.
ix, 195 p. : ill. (some col.) ; 29 cm.

Bibliography: p. 193-195.
Includes index.
ISBN 0-521-25714-X

1. Stars--Color. I. Murdin, Paul. II. Title.

QB816.M34 1984 523.8'2

83-20928
AACR2 MARC

Library of Congress
01699 *67 09249 581322 B © THE BAKER & TAYLOR CO. 5177